Aqueducts and Urbanism in Post-Roman Hispania

Gorgias Studies in Classical and Late Antiquity

26

This series contains monographs and edited volumes on the Greco-Roman world and its transition into Late Antiquity, encompassing political and social structures, knowledge and educational ideals, art, architecture and literature.

Aqueducts and Urbanism in Post-Roman Hispania

Javier Martínez Jiménez

Gorgias Press

2019

Gorgias Press LLC, 954 River Road, Piscataway, NJ, 08854, USA

www.gorgiaspress.com

2019

ISBN 978-1-4632-3915-2

Library of Congress Cataloging-in-Publication Data

A Cataloging-in-Publication Record is available from the Library of Congress.

Printed in the United States of America

TABLE OF CONTENTS

Acknowledgments ix
Preliminary notes xi
Glossary xi
Introduction 1
Chronological and geographical setting 3
Water and cities in the Roman world 7
The first water supply systems 7
Aqueducts 9
The uses of water and the why of aqueducts 10
The transformations of late antiquity 12
Methodology 15
Written sources and earlier research 15
Direct information 19
Relative chronology 20
Scientific dating 23
Indirect information and comparanda 24
Water consuming structures 24
Drains, sewers and latrines 27
Cisterns 28
Late Roman water legislation 29
Comparanda 31
Urban water supply: aqueducts and more beyond the Roman period 35
Quantifying aqueducts in the post-Roman period 35
Aqueducts dated to 400 CE 36
Aqueducts continuing into the post-Roman period 39

Why did aqueducts cease to function? 45
Maintenance costs and expenses..45
Vis maior and unavoidable damage 48
Specialised workmen (or the lack thereof) 50
Late antique alternatives...56
Ground water: Springs and wells 56
Rainwater cisterns 61

Aqueducts, infrastructure and urbanism 67
The persistence of a public water supply 67
Private consumption and domestic architecture 73
Settlement reorganisation 74
Clustering settlements around aqueducts 75
Spreading for new sources 79
Transformations of the hydraulic infrastructure 84
Baths and cisterns 85
Secondary use of conduits 87
The end of aqueducts: the end of Classical urbanism? 88
The emergence of new urban patterns 88
Were aqueducts necessary in the end? 92

Aqueducts and their socio-political importance in Late Antiquity ..95
Aqueducts and the urban elites 95
An elite water supply system? 96
Civic duties 98
Social and cultural redefinitions? 101
The relevance of aqueducts for the Visigothic monarchy....... 103
Germanic monarchies and public works 103
Restoration of public works 105
The aqueduct of *Reccopolis* 107
Water euergetism? 108
The Church and urban water supply systems 109
The new urban leaders 110
Baths and bishops 112
The Umayyad state and the Roman aqueducts 113
The new legal status of the old water supply 113
Looking back, looking East: Syria and al-Andalus 118
Umayyad urbanism and water supply systems up to the end of the Caliphate 121

An aqueduct-less society ... 127

Conclusions ... 129

A surprising degree of continuity? ... 129

The long shadow of the past ... 130

The decline and fall of urban aqueducts ... 132

Further afield? ... 133

Catalogue: functioning aqueducts in the late antique Iberian Peninsula ... 137

Almuñécar ... 137

The remains and course of the aqueduct ... 137

Chronology and dating evidence ... 140

Barcelona ... 141

The increasing political importance of late antique Barcino ... 141

The aqueduct ... 144

Discussion ... 152

Braga ... 153

The aqueduct ... 154

The dating evidence ... 155

Conimbriga ... 156

The aqueduct ... 159

Córdoba ... 161

The early Roman aqueducts ... 163

Late antique water conduits ... 168

Discussion ... 175

Lisbon ... 176

Late antique Lisbon ... 176

The aqueduct and its chronology ... 179

Lugo ... 181

The aqueduct of Lugo ... 182

The chronology ... 185

Mérida ... 186

The water supply ... 188

Dating evidence ... 194

Discussion ... 202

Reccopolis ... 203

The city of *Reccopolis* ... 204

The aqueduct ... 206
Chronology and dating evidence ... 208
Segovia ... 210
The aqueduct ... 211
Discussion ... 215
Seville ... 215
The aqueduct ... 216
An uncertain chronology ... 218
Tarragona ... 222
The multiple water supply systems ... 223
The chronology of the water supply ... 231
The Islamic period ... 234
Discussion ... 235
Valencia ... 236
The aqueduct ... 239
Chronology and dating evidence ... 243
Islamic repairs or evidence for Gothic continuity? ... 249
Zaragoza ... 249
The little known aqueduct ... 250
Dating evidence ... 251
Appendix: Aqueducts and urbanism in post-Roman *Tingitania* ... 253
The aqueducts ... 254
Ceuta ... 256
Late antique urban patterns and water supply ... 261
State formation and water supply ... 262
Conclusions ... 265
Bibliography ... 267
Abbreviations and primary sources ... 267
References ... 270
General Index ... 315
Index of Inscriptions ... 321

Acknowledgments

Arduum videtur res gestas scribere,
Gregory of Tours, DLH, VII.1

This publication written in *The Other Place* is the updated, improved, extended, and yet more concise version of the thesis I defended in 2014 at Oxford. It had in its original acknowledgement section two whole pages: I feel now that an updated, improved, and more concise version would be more fitting.

First and foremost, I have to thank my supervisor, Bryan Ward-Perkins, who has been a constant support and has suffered many early versions of this text. I am still very surprised that he decided to read through all those paragraphs of rather 'inventive' English without sending them back. The great interest he has always shown for my work certainly kept me going chapter after chapter. Similarly, Lukas Schachner, who co-supervised this thesis in its early stages, was always very supportive. Added to them, my two examiners, Chris Wickham and Jim Crow, are also responsible for the final publication of this thesis. Their comments and suggestions have proved essential in giving the book its final form. Alongside these thanks, I should add an apology, for this manuscript was only completed and handed four years after the thesis was defended, although the *muddy* circumstances in which I was may excuse this delay.

Most of my research was carried out in Oxford, where I had the opportunity to discuss my research (and have it amended and corrected) by experts in the subject, both resident Oxonians and visiting academics. Special praise deserve Elena Sánchez, Isaac Sastre, Patricia González and Carlos Tejerizo, with whom I have embarked in many a wild project which have helped consolidating the results of my thesis.

I greatly benefitted from the field work campaigns I carried out at *Reccopolis* with the help of Lauro Olmo and Joaquín Checa from the University of Alcalá, and at Casa Herrera in Mérida, supported by the Consorcio de Mérida and archaeologist extraordinaire Miguel Alba. I had also chances to go to Madrid (at the Instituto Valencia de Don Juan), Valencia, Córdoba, Seville, Barcelona, Princeton, and Vienna, not just to compile data, but also to discuss my interpretations with friends, colleagues and local archaeologists. In the later stages, Andrew Wallace-Hadrill and the rest of the Cambridge 'Impact of the Ancient City' ERC Project[1] have also been very supportive. Despite all of their help, all mistakes and shortcomings are solely my own.

Finally, all my friends in Madrid, Oxford and further beyond have made my long years in Oxford much better. I know they will excuse me if I do not list them all. They deserve most of my non-scientific acknowledgements, but not all; to conclude, and yet even more important, my brothers Arturo and David, my father and my late mother are the ones I really want to thank, because it is because of them that I wrote this.

Oxford/Cambridge, December 2019

---

[1] This project has received funding from the European Research Council (ERC) under the European Union's Horizon 2020 research and innovation programme (grant agreement n° 693418).

Preliminary Notes

Sites, towns and cities will be referred to in their Spanish name for quicker reference, excepting sites such as Lisbon or Seville, or geographical elements like the Tagus or the Pyrenees with overwhelmingly accepted English names. Latin terms and texts will be italicised, as well as texts inserted in other languages (translations provided in inverted commas). Whenever the Peninsula (with a capital P) is referred to, the Iberian Peninsula will be implied.

Unless specified, all figures and translations are the author's own work. All aqueduct maps have been drawn with QGIS on geographic data provided by the Geographical Institute of Spain (IGN); ancient coastlines and roads are based on and modified from the Barrington Atlas, and aqueduct courses derive from my previous research (Sánchez López and Martínez Jiménez 2016).

Before the final manuscript was sent to print, it was brought to my attention by Miquel Martí that the upper courses of the aqueduct of Valencia have been identified and mapped to a source in Tuéjar. This greatly improves our understanding of the aqueduct of Valencia and its source, but should not affect the chronological conclusions presented here.

Glossary

Acequia	Irrigation channel
Alcázar, alcazaba	Castle or palace of Arabic origin
Balneum/balineum, -i	Private/domestic baths
Caput, -itis [aquae]	Source of the aqueduct

Castellum divisorium/aquarum	Terminal/distribution tank
Cuniculus, -i	Tunnel
Domus, -ūs	Urban house
Fons, -tis	Fountain, spring
Hammam	Baths
Nymphaeum, -i	Monumental fountain
Opus, -eris	
caementicium	Mortared rubble
quadratum	Ashlar masonry
testaceum	Brick-faced mortar
sectile	Cut-marble inlaid floor
signinum	Water-proof mortar
Piscina limaria	Settling tank
Saqiya (= acequia)	Conduit or channel
Spiramen, -inis	Manhole
Specus, -ūs	Conduit, usually inside the aqueduct

Matri carae (†2012):

Mente igitur tali tamque alta praedita, inter maritum, gratos filios interque cognatos, omnibus illaesis morienti sensibus, altum obtulit illi animum dederat qui caelitus, ea nunc addat cum coelo gaudiisque repleat almis; nempe sui memores reficit nos mortua heroina.

Adapted from Juan Hurtado de Mendoza (attrib.) 1540 'Hyspana Georgii Manrrici Carmina', *versio Latina* of the 'Coplas' by Jorge Manrique.

Introduction

Aqueducts are, perhaps, the most monumental of the utilitarian works of the Romans, especially when they run on lofty arches. Strabo at the time of Augustus in his *Geographiká* (V.3) praised the conduits of Rome, through which it seemed as if rivers were flowing into the city. This awe at their construction and inherent usefulness is seen later in Cassiodorus (*Var.*, VII.6.3), again when describing the city of Rome. The most famous praise they received was that of Frontinus, who said 'with such an array of indispensable structures carrying so many waters, compare [the aqueducts], if you will, [to] the idle Pyramids or the useless, though famous, works of the Greeks!' (*Aq.*, I.16). Part of this awe towards these structures is preserved in current aqueduct studies. And yet, this aura which surrounds aqueducts as eminently Roman is usually linked to a presupposition that they come to an end with Rome. The following book may change this perception.

Despite the increasing interest on late antique urbanism and the 'transformation' or 'decline' of the Ancient city, the issue of abandonment and reuse of aqueducts has, generally, been left unaddressed. Thirty years ago, Michael Greenhalgh (1989, 110) could claim that there were no archaeological data for the evolution of aqueducts in the post-Roman period, a view which seems to have endured in time.[1] A more recent publication on the archaeology of the West in the late Roman period (Esmonde Cleary 2013) has a very complete index, but lacks any entries for 'aqueduct' or 'water sup-

[1] Although he also seems to have largely ignored Spain in his study.

ply'. Overall, there is a lack of general studies of both aqueducts in the late antique world and comprehensive overviews of archaeological material related to aqueducts and water-related structures. This situation is luckily changing and, slowly, individual aqueducts and regions seem to receive some more attention; a trend which started first for Rome and Italy (Squatriti 1998), and that has now been the Eastern provinces (Pickett 2017). For the Iberian Peninsula or Gaul there has been, to this date, no general approach for the topic.[2] The study of aqueducts and water supply systems may shed new light, and new thoughts, on our understanding of the evolution of Roman urbanism, Roman society, construction techniques, elite culture, and regional economics. Aqueducts are not only an indicator of Roman urbanism, but also symbols of Roman culture, and thus they played an important role in the post-Roman period.

This book is not simply a catalogue of aqueducts and the chronology of their abandonment, although it does include an updated and detailed description of the dating (putting together various types of evidence) of all the sites for which late antique continuity in aqueducts can be inferred. It will be become clear throughout the book that there is indeed much more evidence than what was originally supposed.

The book will be organised into seven chapters, beginning with a chapter describing the methodology used in this study (Ch. 2), and followed by a section on the various systems of water supply available in towns in the period of study (Ch. 3). With this settled, it will be possible to approach the two main interpretative questions that can be addressed by the study of aqueducts (considering both those that continued in use and those that did not), which are the impact they had on urbanism (Ch. 4) and their social and political relevance (Ch. 5). This will be followed by some general conclusions (Ch. 6) and a catalogue with detailed descriptions of the functioning these water conduits in late antique Iberia (Ch. 7). The final appendix gives a study of the aqueducts and urbanism of *Mauritania*

[2] There is, however, and up-to-date catalogue of aqueducts in Spain and Portugal (Sánchez López and Martínez Jiménez 2016).

Tingitana (modern Morocco) which, because of the limited amount of information, it was easier to include separately than in the main body of the text as part of a wider discussion.

But before tackling the main issue of aqueducts, it is necessary to address some general concepts as a way of introduction, especially to the chronological and geographical framework, but also to Roman water culture and the transformations of Late Antiquity. These introductions will serve as a basis for the arguments that will be expanded later and, more generally, to give a broad view on the situation of towns and aqueducts at the end of the Roman period in Iberia.

CHRONOLOGICAL AND GEOGRAPHICAL SETTING

Even if most of the sites mentioned in the text will be well known to readers familiar with Roman archaeology or the Iberian Peninsula, for a broader or less specialised audience it will be necessary to briefly explain the geomorphology and climatology of Spain and Portugal. Similarly, the peninsular periodisation and chronology need some clarification, as the terms used to label periods in Iberian history may not correspond with the general chronology usually linked to them elsewhere in the Mediterranean.

Located in south-western Europe, the Iberian Peninsula is large and square, roughly 750km on each side, linked to the mainland by the stretch of land formed by the Pyrenees. It was formed out of extremely old geological shields, continuously eroded and deeply affected by the Hercinian and Alpine orogenies, which has resulted in its very particular topography (figure 1). As a consequence, the Peninsula is divided into various clearly-defined geographic regions.

The central plateau, or *Meseta*, occupies most of central Spain and it is, on average, 700m a.s.l. It is surrounded on three sides by high Alpine mountains and it gently slopes towards the open west. It is divided into two halves by a mountain range (the Central Range) in a SW-NE axis. Three main rivers cross this plateau from the mountains in the east into Portugal and the ocean: The Tagus, the Duero, and the Guadiana. On the Cantabric (north) coast, the mountains descend very rapidly into the sea and it is heavily truncated by numerous, deep, parallel valleys. This area is also characterised by its abundant vegetation and high rainfalls (figure 2). To the north-

Figure 1: Geographic map of the Iberian Peninsula

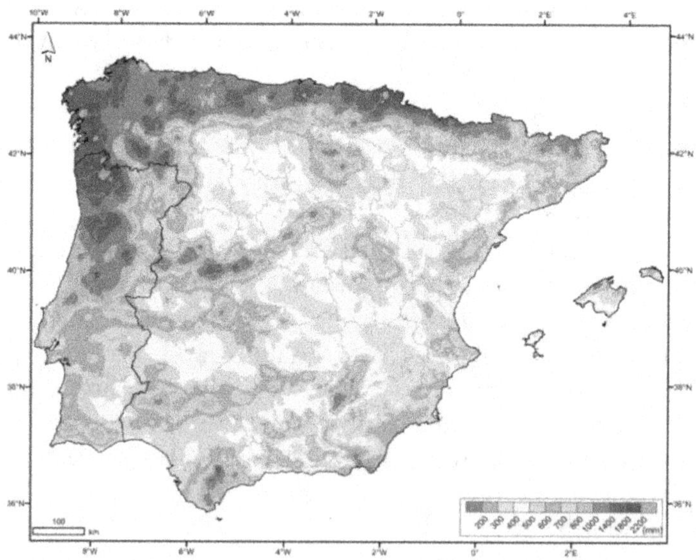

Figure 2: Pluviometric map of the Iberian Peninsula (based on the IM-AEMET 2011 *Iberian Climate Atlas 1971–2000*).

east and the south of the Meseta, beyond the mountains, there are two main river valleys: those of the Ebro (north-east) and the Guadalquivir (south). The Meseta, the Ebro and Guadalquivir valleys, as well as the Mediterranean coast form a macro-region which receives very little rainfall, although the abundance of mountains produces a considerable number of rivers. These are not comparable in size or length to the main European rivers, but they play very important roles on a local scale. Despite this, the great climatic variation between summer and winter usually generates long dry summers, in which droughts are not uncommon.

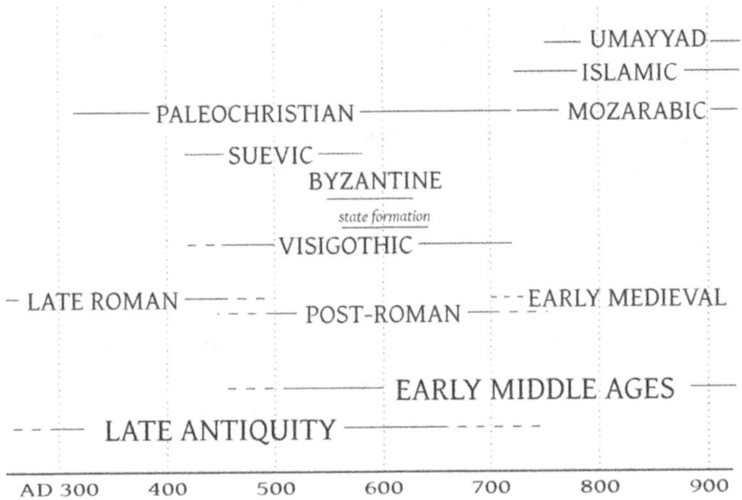

Figure 3: Graph showing the different labels applied to the different political, geographical, and chronological periods for the Iberian Peninsula after the year 300 CE.

In Late Antiquity, the Peninsula (*Hispania*) was divided into six provinces: *Baetica* was roughly the Guadalquivir valley; *Tarraconensis* stretched from Tarragona up to the Cantabrian Range across the Ebro; *Gallaecia* was the mountainous north-west corner; *Lusitania* included the Portuguese lowlands and the western half of the meseta, between the Duero and the Guadiana; *Carthaginensis* formed an odd wedge from the south-eastern Mediterranean coast around Cartagena to the eastern half of the meseta. The Balearic Islands formed a sixth province.

Regarding the chronological terminology, post-Roman periodisation is slightly different from that of the broader Mediterranean (for instance, the Umayyad phase finishes in 1031), but for the purposes of this book we will be using the following labels (figure 3):

- Late Roman, post-Roman, and both early Medieval and Islamic will be used as broad phases, characterised by the material culture of the fourth-fifth, fifth to eighth, and eighth to tenth centuries respectively.
- Visigothic (418-711), Suevic (411–580s), Byzantine (550-630), Asturian (from 720), and Umayyad (from 711) may be used as chrono-political labels (and *not* with an ethnic meaning) within their specific geographic demarcations.
- In a few cases there will be mentions to specific processes of state formation, the most relevant being the Visigothic (570s-630s) and the Umayyad (750s-850s) ones.

The chronological limits of this book are easily described, but defined only with difficulty. The starting date will be 400 CE, just before the arrival of the Germanic invaders. This is a deliberate arbitrary choice: to start with, even if most of the late antique transformations begin already in the fourth century, they are much clearer in the fifth. Furthermore, the changes of the fourth century would require a whole different publication.[3] The abandonment of aqueducts during the fourth century usually happened in small secondary settlements, where it is not possible to identify any of the main long-lasting transformations that characterised the post-Roman period, which further explains why this period has been left out. The arrival of the Germanic invaders in the Peninsula (409) marked the beginning of a period in which Roman cities (which had been going through a long process of evolution already) went under extra social, political and economic stress. This greatly affected urban populations and traditional Roman urbanism.

---

[3] A more general and updated overview on these changes has recently come out (Martínez Jiménez, Sastre de Diego and Tejerizo García 2018).

The closing date is less clear. Ideally, it could be defined as the time when aqueducts finally go out of use. This should be understood as uninterrupted use, although the very interesting transformations of the Umayyad period mean that this period is slightly expanded into the tenth and eleventh centuries. Later medieval reuses and reconstructions after long periods of abandonment will be mentioned, but not taken into full account. These repairs add an extra layer of problems to our understanding of the process of abandonment, because they not only destroyed any trace of late antique use or neglect (especially if these layers were removed to unblock the conduit), but they also make it difficult to distinguish between aqueducts which were in constant use and those that were once out of use in Antiquity and then put back to use.

As far as the geographical limits are concerned, this thesis considers the Iberian Peninsula as a whole, as well as the Balearic Islands. Where possible, other examples from the Languedoc in southern France (*Septimania*) have been considered as well as comparisons, because of the very close political links these regions had in this period. Comparisons and case studies from Morocco (*Tingitania*) have been compiled in the appendix.

WATER AND CITIES IN THE ROMAN WORLD

While aqueducts are the most easily recognisable elements of Roman water supply systems, they were not simply supplying drinking water, as it can be assumed from a modern perspective, nor were they the only ways in which cities and towns obtained water.

The first water supply systems

The dense habitation that characterises urban settlements has always made water supply a matter of concern for town dwellers – water is essential for life and, the more concentrated the population, the higher the stress on local hydric resources.

Despite some badly founded claims that Sagunto had an aqueduct built by the Iberians before the Second Punic War (Civera Gómez 2004, 2008), and the existence of pre-Roman irrigation channels in Marroquíes Bajos and Martos (Sánchez López and Gozalbes Cravioto 2012), long distance urban water supplies arrived in the Peninsula with the Romans. Springs, wells, rainwater cisterns, and rivers were the only pre-aqueduct sources of water in towns; they continued to be so while aqueducts functioned, and after the aqueducts fell out of use. This is a common feature of all the urban or proto-urban settlements of pre-Roman Spain, as well as of the first Roman republican settlements.

Springs are places where water flows naturally out of the ground, and Frontinus mentions them as one of the main sources of water in the city of Rome, especially because they were healthy and even curative (*Aq.*, I.4). These springs also supplied the inhabitants of Rome after the aqueducts were cut during the Ostrogothic siege of 537 (*Lib. Pont.*, LX.4.5). Intra-mural and peri-urban springs seem also to have been very common in Spanish towns, and even though there is little positive evidence for their use in Late Antiquity, it is very probable that they were used alongside aqueducts, and even more once these ceased to function.

Wells (Hodge 1992, 50-3) are pits dug into the soil that reach the water table, and despite the procedures mentioned by Vitruvius (*Arch.*, VIII.1.1–7) on how to find a good spot, it seems that well-digging was highly speculative. Because of their nature, in which water needs to be pulled out of the well, well-shafts need to be lined (with masonry, brickwork and even old barrels), lest they collapsed, and must have some sort of support to allow the rope to be pulled; all of which should make wells clearly visible in the archaeological record. In purely archaeological terms, wells are negative structures, so they can easily be given relative chronologies based on the contexts they are cut into, and their abandonment is dated by the infill.

Regarding rainwater cisterns, they appear usually (but not exclusively) in dwelling contexts, so each domestic unit would have had one to suffice its water needs – or to complement the water obtained from wells and springs. In fact, because they can be refilled on a regular basis, offer a large reservoir, and are easily built, cisterns could be seen as the main type of water supply system until aqueducts developed (Castro García 2017). Furthermore, once aqueducts were built, cisterns could be used to store aqueduct water as well. Pre-Roman cities such as *Lucentum* (figure 4), *Bilbilis* or Ampurias relied on cistern supply all throughout their history, and even Roman foundations like Medina Sidonia had extensive cistern networks. The most complicated of these even tapped water into the cisterns directly from the water table, and from this it could be used elsewhere in the city.

Figure 4: Roman urban cistern lined in opus signinum from the site of Lucentum, near Alicante.

Aqueducts

Aqueducts (*aquae ductus*, 'water conduit') are long-distance water supply systems, developed in Antiquity in various urban cultures, wherever water supply became an issue (Wilson 2012, Hodge 1989), although the Romans were responsible for their expansion across the Mediterranean.

The Romans mastered their construction and spread their use at the pace of the Romanisation, especially after the development of vaulted structures and pozzolanic mortars in the last centuries BCE. The principles of aqueduct functioning are very simple: Aqueducts tap water from a source (a dam, a spring, a river), which is usually referred to as the *caput aquae*. From there, a conduit (the *specus*) is built in masonry walls (*substructio*), cut in the rock, or dug into the soil, leading the water to its final destination, the *castellum aquae* or *castellum divisorium*, from where the water was redistributed throughout the city.

Because the conduit has a constant slope with a pre-calculated gentle gradient (an element which required very precise calculations and engineering), most conduits are built following terrain contour lines, but when it becomes necessary to keep the gradient over the ground for a long distance, normally sections elevated on arches (*arcuationes*) would be built. Even if these are the most famous 'aqueducts' (like the Pont du Gard), they correspond to a minimal percentage of an aqueduct's course. Along the way there may be settling tanks (*piscinae limariae*) that would be cleaned regularly from any debris that might have been carried by the water. The conduit is almost always covered (with slabs, tiles or vaults), to protect water from the light and the elements, whereas the conduit itself is lined so to make it water-tight either with clay or *opus signinum*.

Aqueducts usually run on gravity flow but, whenever necessary, the Romans could build pressurised conduits in lead, stone or clay pipes. This usually was done in order to cross deep valleys without building high bridges. This solution is called an inverted siphon, because it functions with the basic hydrostatic principles by which gravity pushes water into the pressurised conduit, and by atmospheric pressure, it emerges on the exit side at the same height as the intake.

The uses of water and the why of aqueducts

With the development of a more complex urbanism in the Roman period, the demand for water in urban contexts increased. While in cities there is always a need for clean drinking water (and the greater the population, the bigger the stress on the available water resources), in urban contexts it was not the only demand. The high population density and the intense economic activity inside the city and its surroundings meant that what cities demanded was primarily quantity and, ideally, quality (but not exclusively). Whereas domestic drinking and cooking water could be easily supplied from springs or cisterns (the most desirable water: *Arch.*, VIII.2-3), it would seem a waste to use this water for workshops, construction, washing, cleaning, irrigation, gardening, flushing sewers or bathing.[4] That is why water supply needed to maximise its resources, and an integrated supply system with aqueducts, cisterns, wells and rivers provided with all the necessities (Laurence, Esmonde Cleary and Sears 2011, 23-30).

Furthermore, in Roman contexts public water supply was also politically and culturally important, because public baths and fountains, copying Italic models, were symbols of *Romanitas* (Hodge 2000a, Yegül 1991). Aqueducts played a major role in this context of underlining the Romanness of its citizens *and* multiplying the available volume of water available for public and private use. The first aqueducts are to be found in Augustan foundations or refoundations such as Mérida (*Emerita Augusta*), Zaragoza (*Caesaraugusta*) or Tarragona (*Tarraco*), or cities granted special privileges during the Republic, like Cartagena (*Carthago Nova*). These cities were not only administrative and economic centres, but also local models of the new Roman urbanism spreading across the Mediterranean. These new foundations were sometimes equipped with aqueducts from their very beginnings, which removed immediate water availability as a constraint when choosing the location for a new settlement (Hodge 1989, 128-9, J. B. Ward-Perkins 1974, 3-4).

An element of this Roman urban culture was public baths, which required large quantities of water (Laurence, Esmonde Cleary and Sears 2011, 203-4, 215). In Spain, the earliest baths known are the

---

[4] As modern society does nowadays, where we use drinking water to flush the toilet.

forum baths of Valencia, which are Republican in date (Marín Jordá and Ribera Lacomba 2000). Many other Roman baths are known from the second century BCE to the sixth CE, mostly private, but also various public ones, both rural and urban (García-Entero 2005), although in Spain there are no large public baths of imperial scale, as those of Rome, Carthage, Arles, or Constantinople.

During the second half of the first century CE, the Iberian Peninsula was fully integrated in the Roman imperial political system (Houten 2018) and Mediterranean urbanism and Italic bathing culture were firmly established. Furthermore, by means of the Edict of Vespasian, towns in *Hispania* acquired Latin rights, thus blurring the juridical status between native settlements and Roman towns. This led to an increasing interest in the urban elites to demonstrate their *Romanitas* and it is in this period when most of the monumental city centres of Roman towns in Spain were erected. Inter-town competition, and peer-polity interaction prompted the development of large forums, basilicas, baths, and aqueducts – although the chronological correlation is not enough to claim direct causality links (Martínez Jiménez 2019).

THE TRANSFORMATIONS OF LATE ANTIQUITY

After the first centuries of Roman urbanisation, towns in *Hispania* and *Tingitania* were not only heavily monumentalised, but most of them were equipped with aqueducts, and it is safe to say that all of them had a set of baths. This would cease to be the norm after the fourth century, after which there were a series of transformations which define the end of the Roman world. These changes can be, very broadly, simplified into political and administrative, economic, and cultural and social (Martínez Jiménez, Sastre de Diego and Tejerizo García 2018).

From a political and administrative point of view, the early Roman curial administration, which had been responsible for the construction of aqueducts in the first place, was thoroughly transformed. The early Roman model was at first substituted by the late Roman one, as a result from the Diocletianic-Constantinian reforms of the early fourth century (Liebeschuetz 1992, Esmonde Cleary 2013, Curchin 2014). This evolved during the late antique centuries with only minor modifications in the Visigothic period, which was then substituted by an Islamic, Umayyad one. These political transitions had a direct effect on townscapes and urbanism, because each administration developed a new architecture of power (with changing monumental priorities), and (at a higher level) new hierarchies of cities, promoting some while relegating others. These changes do not reflect long-lasting periods of political instability, but rather short moments of 'state formation' followed by decades of application, development of preservation of the new model. The former can usually be seen through the dismantling of the monuments of the old regime, and the construction of the new ones that represent and identify the new administration. There are also moments of power vacuum and lack of political activity, but these were, largely, times of stagnation in terms of urban construction (Martínez Jiménez 2013).

Economically, there are two main transformations. The first one is, in those areas still under imperial control, the reorganisation of centralised taxation as a result of the fourth century administrative reform. The new tax regime affected the amount of funds available for monument upkeep, as more money had to be diverted to maintain the imperial army and the centralised bureaucracy. The changing political incentives and taxes also limited the amount of private funds which were devoted to public munificence. The second one was the end of the unified imperial system articulated around Mediterranean and Atlantic trade, which shifted to stronger regional and local patterns after the fifth century (Wickham 2005). These economic circumstances, together with the changing political situation(s) mentioned earlier can help to explain why Roman townscapes (and their water supplies) changed. But there were also very important cultural changes to take into account.

The evolution of Roman society after the third century has to consider, first of all, the Christianisation of the elites and of the Roman state as a whole. The Church became a main force with its own monumental schemes and ideas on some old Roman buildings. The elites were not only Christianised, but also militarised, especially after the integration of the Germanic peoples into the Roman provincial system. The approach the changing elites had to Roman water supplies in particular and urban monuments as a whole was as much a result of the political-economical framework as of the new cultural priorities. Similarly, the arrival in the Peninsula of the Umayyads added an extra layer of cultural and religious complexity. Furthermore, this new cultural component was a new ruling elite which did not link itself with the Roman past and had, at best, a neutral approach towards past monuments. This contrasts sharply with the Christian, militarised elites of the immediate post-Roman centuries, which still saw Rome as a cultural idea to which they could belong.

***

Overall, aqueducts were too massive and too useful to be consciously pulled down without a specific reason, but their functional continuity is quite a different story. As with many other Roman buildings, they went through a process of slow decline, which is roughly covered by the chronological framework of this book. Still, the reasons for this slow decline are intriguing.

In order to identify these reasons, it is necessary to look at the archaeological evidence for aqueducts and their urban contexts, not only in chronological sequence, but also from a diachronic perspective. The social and political context will be equally important, because the transformation of Roman society is deeply interdependent with the evolution of urban townscapes and infrastructure. Aqueducts were elements of the Roman city, built with a specific purpose, in a specific context and for a specific audience, and their continuity is the result of various local responses to systemic changes. Late Antiquity was, after all, a period of transformation away from the Roman period, a transformation of the Roman world, and the role of aqueducts in this new context had to be renegotiated.

METHODOLOGY

A necessary step before commencing the analysis of the archaeological evidence is to explain the methodology followed and the different approaches taken. This is also the best way to explain the questions that may arise, and how they can be answered, especially for those cases when the evidence is not as solid as one could hope for and when cautious extrapolations need to be made. This chapter will first address the information that can be obtained from contemporary and secondary sources, followed by an account on the various useful archaeological approaches and their methodological shortcomings.

WRITTEN SOURCES AND EARLIER RESEARCH

From our modern perspective, aqueducts are seen great monuments, but outside Rome (where they were constantly praised by ancient authors) their monumentality seems to have been eclipsed by their tremendous functionality. They hardly appear in texts from the provinces, just as modern literature only mentions water pipes if necessary and incidentally.[1]

In the late Roman and Visigothic period at least, none of the written sources mention or describe aqueducts. They sometimes mention elements of urbanism or the construction of churches, but not much else. Chronicles (Hydatius' and John of Biclar's chronicles,

[1] Those ancient writers who wrote about aqueducts had a clear agenda behind them: Strabo was dedicating his work to Augustus, Frontinus (*Aq.*, I.16) was the *curator aquarum*, and Cassiodorus (*Var.*, VII.6.2) praises them in the formula in which the 'count of the aqueducts' is instituted.

the anonymous *Chronica Caesaraugustana*, and Isidore's *Chronica* and the *Historia Gothorum*) tend to be quite schematic. Hagiographies give only circumstantial evidence, although the *Vitae Sanctorum Patrum Emeritensium* give a very good picture of sixth-century Mérida. The *Historia Wambae Regis* by Julian of Toledo is a short account of a military campaign. Visigothic laws, most of which collected older Roman legislation still in use, do not mention aqueducts; They only regulate the use of irrigation channels in rural areas (*LV* VIII.4.31: *De furantibus aquas ex decursibus alienis*). National church councils are not helpful either, and Isidore's *magnum opus*, his *Etymologies*, only mentions baths as urban buildings, and water pipes as parts of the house.[2] Isidore's mention of pipes as part of the house shows that he was familiar with piped water in urban domestic contexts, although how much his knowledge derives from the writings of Vitruvius or Frontinus (such as the fact that pipes are classified according to their capacity) and how much from his own personal experience is not known.

In the Islamic period, however, we are luckier, as emirs and caliphs recorded their works on aqueducts in public inscriptions. These are also recorded in the chronicles, such as the *Chronicle of al-Nasir*, the *Akhbar Maymuᶜa* or the *Bayan* by Ibn ᶜIdari. Islamic geographers such as al-Idrisi, al-Bakri and al-Himyari were very interested in Roman monuments, including aqueducts, although by this time most of them were certainly out of use, as they are described as abandoned ruins. The tradition of Arabic geographies and description of monuments continued well into the early Modern period. Latin and Romance writers of the medieval period mention the aqueduct of Segovia (as Rodrigo Ximénez de Rada, in his *De Rebus Hispaniae*) and that of Barcelona, which must have been a preeminent monu-

[2] *Balnea* and *thermae*: XV.2.39-40. Pipes: XV.8.17; although the mention of pipes seems to refer back to the works of Frontinus, mentioning how pipes are classified according to their capacity: *Fistulae aquarum sunt dictae quod aquas fundant et mittant; nam στολα Graece 'mittere' est. Formae earum pro magnitudine aquae et capacitatis modo fiunt* = Water pipes are called this because they carry water, for 'stola' is 'carry' in Greek. Their different types are classified according to the amount of water [that flows in them].

ment in the landscape, as it is used constantly as a reference in charters (Mayer Olivé and Rodà de Llanza 1977), but these mentions are often incidental.

The first modern enquiries about Roman aqueducts seem to have been carried out by order of Queen Isabella I (r. 1474-1504), who commissioned the restoration of the aqueduct of Segovia (Fernández Casado 2008 [1972], 65).[3] From the Modern period, there are several studies of the Roman monuments of Spain, like Pons D'Icart's *Grandezas de Tarragona* (1572). The most extensive study of an aqueduct was made by Andrés Gómez de Somorrostro, who wrote his *El acueducto y otras antigüedades de Segovia* in 1820, which included a complete study of its course, its architecture and its design (until then all previous studies were largely descriptive). In a similar fashion, Pascual Madoz published his *Diccionario Geográfico* in 1849, in which he described most of the aqueducts still standing in his day. This tradition of architectural description of the aqueducts continued well into the twentieth century.

The first true archaeological studies of the aqueducts in Spain were carried out not by an archaeologist, but by a civil engineer, Dr Carlos Fernández Casado, who was the provincial engineer of Granada. In the early 1930s he began to record and to study Roman infrastructures, not only aqueducts, but also roads and bridges.[4] His survey of the aqueduct of Almuñécar was so precise that recent archaeological surveys confirm his proposed course with only a very small margin of error (Sánchez López and Moreno Pérez 2012). After the Civil War (1936-39) he was deprived of his title of engineer and his right to teach, so he began to study history, law, and geography. After obtaining his second degree he was given back his title of engineer and was teaching in the University in Madrid while he obtained his doctorate in civil engineering. Carlos Fernández Casado is responsible for the first extensive publication on Roman aqueducts in

[3] Although previous medieval repairs are possible – see below.

[4] I have to thank Carlos Fernández Casado's grandson, Javier Muñoz-Rojas Fernández, for his help in compiling the information published by Fernández Casado, and for facilitating my access to his archive.

Spain, in which he put together all the information he had compiled in his early years as an engineer (Fernández Casado 2008 [1972]). The fully-illustrated publication gives a detailed account of the engineering, the architecture and (to an extent) the archaeology of the monumental aqueducts of Spain, including written references to the structures. The main shortcoming with his work is that, because of the nature of his publication (a study of Roman engineering), he only considered aqueducts to be the arched structures, leaving aside or barely mentioning underground conduits. This explains why he did not include in his publication the works on the aqueduct of Valencia (Fletcher Valls 1958). Maybe by coincidence, but all the archaeological publications of the main aqueducts of the Peninsula post-date 1972, when Fernández Casado published his work.[5] In the early twenty-first century, aqueducts have been studied again, revising old publications and introducing the contributions of urban rescue archaeology and GIS modelling, greatly improving our knowledge of those aqueducts belonging to continuously-occupied cities. This led to the 2002 monographic edition of the journal *Empúries*, two congresses in 2009 on ancient hydraulic technology (*Aquam perducendam curavit* held in Cádiz and published in 2010, and *Wasserversorgung in Toledo und Wissensvermittlung von der Antike ins Mittelalter*, published in 2017), and the 2011 publication on Roman sewers edited by Josep Anton Remolà and Jesús Acero. Lastly, it should be noted that there has been in the recent years a number of doctoral theses specifically focused on water, including those of Elena Sánchez about Almuñécar, Mário Soares Fortes on Portugal, and of Jenny Pérez and Macarena Lara Medina about Cádiz, besides my own.

---

[5] Segovia (Almagro Gorbea and Caballero Zoreda 1977, Ramírez Gallardo 1975); Mérida (Álvarez Martínez 1977); Tiermes (Argente Oliver and Díaz Díaz 1980); *Italica* (Canto 1978); *Conimbriga* (Étienne and Alarcão 1974); Baelo Claudia (Jiménez 1974); Barcelona (Mayer Olivé and Rodà de Llanza 1977); Tarragona (Sáenz Ridruejo 1977, Cortés, Benet and Bermúdez 1989).

DIRECT INFORMATION

The direct archaeological excavation of aqueducts is rare, and most of the early studies carried out were done on the over-ground arched sections, which only constitute a small portion of their total length. This means that there is little direct archaeological information obtained from aqueducts, which is key in order to get a correct idea of the chronology of use and abandonment.

A precise assessment of the available chronological evidence is extremely essential in this current discussion, not only because vague terms such as 'late antique' or 'early medieval' could mean anything between the fourth and the seventh centuries, but also because there is a great difference between terms usually found such as 'fourth to fifth century', 'late fourth' and 'early fifth'. Nevertheless, it has to be acknowledged that the use of these concepts is sometimes unavoidable, due to the nature of the evidence, but the objective of this research is to obtain a result as accurate as possible.[6] This is obtained through a careful examination of the archaeological material, as the written sources of this period are hardly of any help. Besides that, there is the added problem of dating (and identifying) abandonment contexts which is, in archaeological terms, a difficult task.

This is always complicated, as these need to be carefully registered and excavated, which sadly in Spain and Portugal has only been the case in the past thirty years. On the plus side, in those thirty years there has been an explosion in late antique archaeology, and with its increasing popularity, ceramic typologies (which are the essential key to date most archaeological contexts) have become more accurate and precise (Wood and Martínez Jiménez 2016). This does not only apply to imported materials, but also to local productions, especially thanks to the excavations of sites like *Reccopolis*, the theatre of Carta-

---

[6] Canto 1978: 336, for instance. This happens in Gaulish examples as well: Didier Rigal (1997, 313) mentions for the aqueduct of Cahors: '*la date d'abandon est très envisageable au vu d'autres aqueducs gallo-romains* (...)' = 'the abandonment date is very predictable based on [those of] other Gallo-Roman aqueducts'.

gena or El Tolmo de Minateda (Martínez Jiménez, Sastre de Diego and Tejerizo García 2018).

Obviously, chronological data obtained directly from the aqueducts would be in all cases preferable, because it provides primary dating information on repairs, maintenance, destruction, abandonment or continuity of the conduit. Alas, because of the nature of the conduits (closed and usually vaulted), dating material (mostly pottery) made its way into aqueducts only with great difficulty, so indirect information about water-consuming structures will also be taken into consideration, because this will be the largest part of the available corpus of data.

Relative chronology

Well-dated contexts excavated from aqueducts are the most useful way to create a relative chronology, as they provide either a *terminus ante quem* or a *terminus post quem* for the abandonment. Other factors, such as the nature of the context (if it was sealed, or closed, or if it is an abandonment layer, or deliberate backfill) also have much to say about the reliability of the relative dating. In this way, the coin of Trajan found inside the bottom *opus signinum* layer of the *specus* of the aqueduct of Segovia gives a *t.p.q* for the lining of the *specus*, for instance (Zamora Canellada 2007). And the presence of a dump in a terminal cistern of one of the aqueducts of Córdoba provides a *t.a.q.* for the abandonment of the aqueduct (Carmona Berenguer, Moreno Almenara and González Viseda 2008).

Other elements providing relative chronologies are restorations, repairs, and other elements that may indicate continued (or secondary) use. Some Iberian aqueducts present restorations even if this is more widely known for examples outside Spain, like the brick reinforcements of the aqueducts of Rome (Martínez Jiménez 2010) or the reconstructions of Arles and Nîmes (Fabre, Fiches and Paillet 1991a, 1991b, Raffard, et al. 2000). The only way to date these restorations is normally by means of their construction technique, which is not necessarily a good indicator, as the dates are usually estimates. For example, the buttresses placed on the sides of the pillars of one of the aqueducts of Mérida were built in a different construction technique, which would only indicate that the aqueducts were repaired at a later date (figure 5). Generally, for this period it is assumed that the coarser the technique, the later its date. Despite being very 'catastrophist', this assumes a decline and loss of technical abilities (or simply

Figure 5: Butress reinforcing the pillars of the bridge over the Albarregas river, part of the Proserpina (also known as Los Milagros) aqueduct in Mérida

of skilled labour).[7] This still has some truth behind it at least as far as hydraulics are concerned: the increasing size of *caementa* in the *opus signinum* is a constant from the second into the fourth centuries (Lamprecht 1987), and the Visigothic *opus signinum* from *Reccopolis*, Mérida and Toledo seems to have been made with chunks of tiles and pots that were not even properly crushed.[8] Such arguments about the date of different construction techniques has led to intense debates about the date of the dams of Proserpina and Cornalvo outside Mérida, traditionally considered to be Roman, and the problem is still unresolved (Feijoo Martínez 2005). In a few cases the use of earlier material or *spolia* in later constructions provides a *post quem* for these, as with the Roman column drums moved from the curia of the forum of Valencia (which was still standing in the late Visigothic period) and used to build a new Islamic water conduit (Martí and Pascual 2000).

In some cases, it is possible to give relative dates on the basis of the new technological developments or the abandonment of old Roman practices in the Islamic period. The clearest examples are the abandonment of *opus signinum* as lining for conduits and cisterns in favour of water-proof red stucco (Ventura Villanueva 1993, 33-4). The introduction of this technique can be dated to the late emirate and the early caliphate, in the early tenth century.[9] Despite this, the Islamic wide knowledge of hydraulic engineering meant that many aqueducts were reused and, as mentioned above, this was often done after removing all archaeological traces of late antique evidence (abandonment layers, sinter deposits, repairs, etc.).

Finally, pipes should be mentioned, even if they hardly survive in the archaeological record. The fact that they were usually made out of lead means that they were an easy target for theft and recycling, and the material obtained from the spoliation trench indicates

[7] As discussed by Ward-Perkins (1997, 2005, 104-10).

[8] Cf. the typologies in Palestine made by Porath (2002).

[9] However, Umayyad aqueducts in the Levant in the seventh and early eighth century are built with *opus signinum* linings (Porath 2002, 35), indicating that the stucco lining may be in fact a new development.

a probable *terminus ante quem* for the abandonment of the aqueduct, because that should predate the removal of the pipes.

Scientific dating

Only in few cases it is possible to obtain an absolute chronology to date the construction/abandonment/destruction of an aqueduct, largely because of the nature of the dating methods. Radiocarbon, thermoluminiscence (TL) or electron spin resonance (ESR) may be used to date abandonment of aqueducts in an absolute sense with various degrees of accuracy. Although there are hardly any organic remains to be found in aqueducts, AMS-radiocarbon can still be used to date lime mortars (Alejandre Sánchez 2018), which could be used to calculate the dates of construction and of later building phases, which could be extremely useful (this has not been applied to aqueducts yet).

Sinter is a precipitation of the calcium carbonates ($CaCO_3$) dissolved in the water (Sürmelihindi, et al. 2013). Its precipitation rate is not universal or constant, because it depends on the type of water, the gradient of the conduit, seasonal flows and the action of microorganisms (stromatolites), which accelerate the deposition rate (Fabre, Fiches and Paillet 1991a, 172-9). For this reason, the precipitation rate varies between 0.0123 to 44 mm/year, although the average seems to be 1mm/year (Fabre, Fiches and Paillet 1991a, 191, Hodge 2000a, 60-1, Keenan-Jones, Hellstrom and Drysdale 2008, 331, Sürmelihindi, et al. 2013), and deposition could also be microscopical (Passchier, et al. 2016). For approximate estimates, the average rate can be used as a rough guide to date sinter deposits. In this way, the thickness of the concretion indicates for how long water was flowing through the conduit since it was last cleaned.

However, the formation of sinter can also be dated scientifically with the combined application of ESR and Uranium/Thorium series which can, potentially, give a date with a precision range of 2-10% (Blackwell 2006). With proper sampling method, ESR could date the formation of the last layer of sinter, giving a date range for the last flow of water inside the conduit. This is, however, only a theoretical proposal which has not carried out yet for aqueducts (M. Duval, et al. 2012, Keenan-Jones, Hellstrom and Drysdale 2008, Sürmelihindi, et al. 2013).

INDIRECT INFORMATION AND COMPARANDA

Whereas the dating obtained directly from an aqueduct is the most desirable and the most specific, we often have to rely on water-related structures to infer the continuity of the aqueduct. The problem with this approach is that the link between aqueducts and water consuming structures cannot be taken for granted, when it is perfectly possible that the water consuming buildings (baths, fountains) functioned without an aqueduct supply system. The real difference between the direct information mentioned above and the indirect information below is that the latter cannot be used independently to date the continuity/end of the aqueduct (whilst the former certainly can). Thus it is necessary to use the different pieces of indirect information together in order to make a coherent statement of continuity, although ideally the combination of both types of evidence will give the most accurate answer to the question of dating the use of the aqueduct.

Water consuming structures

The most evident water consuming structures, and those most visible in the archaeological record, are baths, baptisteries, and (to a lesser extent) fountains. Furthermore, as these structures are notable features of Roman and late Roman urbanism, they not only provide dating evidence for the aqueduct, but also for distribution patterns and social use of water.

Baths were a key element in Roman towns – and they continued to be so in Late Antiquity. In the sixth century for instance, during the Gothic wars, the main concern of the citizens of Rome once the aqueducts were cut was that they could not take baths anymore, because these were fed by the aqueducts (*Bell. Goth.*, I.19.28 and I.20.5).[10] Even if during the late antique period bathing had to face the partial opposition of the Church (which opposed the pagan connotations of bathing and imposed new concepts on what was luxurious and what was needed, further discussed below) many baths sur-

[10] The authorities were, however, more concerned with the powering of the flour mills, which had to be relocated to the Tiber (*Bell. Goth.*, I.19.20-6).

vived the fourth century, and continued well into the fifth. This is evident because baths are almost constantly in need of repair and maintenance. Baths, in fact, required extensive resources for their functioning and repair, and not just water, but especially fuel (Blyth 1999, 90-1, Meiggs 1982, 258) (cf. *CTheod* 13.5.10 and 14.5.1). These were difficult to achieve in Late Antiquity, but still was done because baths were elements of civic pride (Yegül 1991, 321–3). Whenever a bath continues to function into the fifth century or later, it is probable that it was supplied water by an aqueduct. With new baths built in this period, it is also possible to assume a similar situation. Overall, unless a clear alternative water-supply system is identifiable, as it happens with the sixth-century baths of Tarragona which were supplied by a well (Remolà Vallverdú and Ruiz de Arbulo 2002, 63), baths are good indicators of continuing aqueducts. Likewise, any modification to the bath design or infrastructure gives further information about the use and availability of water: baths seem to have reduced or abandoned their hot rooms in Late Antiquity (DeLaine 1997, 39-40, Nielsen 1990, 57), which can respond both to fuel shortage and/or water scarcity.

Baptisteries are a different issue, because they certainly require water but not a permanent supply, especially because in this period baptism was only done once a year at Easter. In theory these structures do not need piped water, even though there are examples in Gaul where this was the case, like in Lyon, Poitiers, Aix-en-Provence and Venasque (Benoit and Rouillard 2000, 168, Grewe 1991, 19-25). In the Spanish examples there is no direct evidence to link piped water with baptisteries, but their presence close to other water-consuming structures, like a fountain in Valencia (Ribera Lacomba 2005) or a bath complex in Barcelona (Bonnet and Beltrán de Heredia 2002), can be considered as a possible indicator of a functioning aqueduct.

Fountains can only work if there is running water. Sometimes fountains are dedicated with an inscription, which tends to date their construction and indicates a time in which the aqueduct was still in use, even though late fountains are rare (only one inscription possibly referring to a fountain is known in Spain for this period: *CIL* II 4109 from Tarragona). Archaeological dating for repairs/maintenance of fountains can be linked to a functioning aqueduct with a high degree of certainty.

Lastly, workshops such as fulleries, potteries, salteries, dyers, glass and metal workshops many times required running water for either cleaning the facilities, powering machinery or even actually in the final product. Even if Roman legislation prohibited the presence of workshops inside cities, these may have been located in the suburbs, and could have benefitted from the presence of aqueducts.[11]

The most common type of workshop complex in coastal towns were fish processing factories, the *cetariae* where *garum*, salted fish, and other fish-derived products were prepared.[12] In these, water was needed for the elaboration of some products, and also for boiling the guts, cleaning the fish, the vats, and the employees (Wilson 1997: 180). The abandonment of a factory gives a rather unreliable *terminus ante quem* for the abandonment of the aqueduct: the end of the aqueduct could cause the end of the fish factory although alternative water supplies, like cisterns, might have been more common (Sánchez López 2018). This may have been the case with the aqueduct of *Baelo Claudia*, which was destroyed by an earthquake, soon followed by the abandonment of the fish factories (Jiménez 1974, 293, Lagóstena Barrios 2001, 124-5). This, however, is not necessarily a *sequitur*, as the aqueduct could potentially have continued functioning after the end of the fish processing workshops, supplying the city.

There are not many other known industrial trades inside Spanish Roman towns, or at least not through their archaeological remains. Some olive and wine presses are known in Valencia (Álvarez García, Ballester Martínez, et al. 2005) and Barcelona (Beltrán de Heredia 2002b) for the earliest part of the period considered in this work, but these do not directly need a constant source of water, like

---

[11] Miko Flohr (2006) and Elena Sánchez (2018) have been working on these issues in the last years, but much still needs to be done.

[12] Particularly in the late period, as only those fish factories that were located in urban or suburban areas survived in the fifth century (Lagóstena Barrios 2001, 355-6).

tanners, potters, dyers or fullers. The only other main complexes intra-muros from the Visigothic period are the glass and gold workshops of *Reccopolis*, located outside the palace complex (Castro Priego and Gómez de la Torre-Verdejo 2008), but these do not provide any concrete dating evidence linked to its aqueduct.

Drains, sewers and latrines

Because of the nature of Roman hydraulics, water was constantly flowing through pipes and conduits, as if the aqueduct flow were to be stopped, the water would overflow in the conduits, damaging the structure (Hodge 2000a, 48). That is why drains and sewers were essential in cities with aqueducts, so water could be evacuated away from the streets. Therefore, and in theory, non-functional or collapsed sewers would indicate that the aqueduct was out of use about the time in which the sewers were not in use. However, it would be too risky to assign a direct causal relationship between the end of functional sewers and the end of functional aqueducts, especially because drains can survive after the end of the aqueduct, and alternatives (over-ground gutters) are possible.

Aqueduct overflow is not the only source of water for sewers, and in fact sewers may have originated as ways to prevent flooding or to limit the risk of damage caused by rain-water, so the construction of new drains, like those of Cartagena cannot be linked to a functional aqueduct (Egea Vivancos 2002, Egea Vivancos, Ruiz Valderas and Vizcaíno Sánchez 2011).

The knowledge of drains in Roman cities in Spain is, alas, limited to a few sites, and even amongst these the degree of knowledge is very uneven.[13] The evidence for the continuity of sewer systems is not plentiful, and only in a few cases late antique levels are known. Even if many sewer systems seem to be abandoned or appear to have collapsed at some point in Late Antiquity, it is equally true that

---

[13] Remolà Vallverdú and Acero Pérez (eds.) 2011 contains papers on fourteen different sites, including Córdoba, Barcelona, Valencia, Seville, Lisbon, Zaragoza, Valencia and Cartagena. Two updates have been recently published (Ruiz Bueno 2018a, Acero Pérez 2018).

many Roman drains were constantly maintained and in use – especially in the Umayyad period, but some even into our day. These long-lasting sewers are, mostly, those which were so big that could only be clogged by a major collapse of the vault, rather than as a result of poor maintenance. These were usually kept clean by rainwater flow (Ruiz Bueno 2018a, 147). This constant cleaning has probably destroyed any stratigraphic information about late antique use or abandonment. In any case, it is possible to identify several new drains built after the fifth century in Barcelona, Seville and Cartagena.[14]

Cisterns

Cisterns have been presented already as examples of (potentially) alternative supplies to aqueduct water. However, in some circumstances these can be seen as elements used in combination with aqueduct supply. Their presence in late antique contexts has normally been taken as a sign of an abandoned aqueduct, as it is assumed that this system is linked to the collection of rainwater and substituted the lack of piped water (Bosch Puche, et al. 2005). However, I believe that, in some cases, cisterns could indicate a functioning aqueduct, and it is clear that cisterns and aqueduct coexisted in the Roman period (Castro García 2017, El Khatib-Boujibar 1992). They would indicate the need to create a domestic water reservoir, perhaps linked to a seasonal flow or to a limited or unreliable conduit. Alternatively, cisterns could have been used as ways of storing water and of maintaining the hydraulic pressure inside pipes. Cisterns also stored water during the night, when water consumption is virtually none, thus collecting the constantly flowing water so it could be used during the day. This is essential in sites where the flow was seasonal or reduced (Brunn 2002b, 219-20). They thus fulfil the function of terminal tanks, but on a smaller scale, and there are various examples in the Roman world that support this idea. In Rome, huge cisterns were

[14] These are issues which have been addressed in general (Acero Pérez 2011, 163) and specifically for Barcelona (Beltrán de Heredia and Carreras Monfort 2011), Cartagena (Egea Vivancos 2002), and Seville (Jiménez Sancho 2011).

added at the terminal deposit of the *Aqua Antoniniana*, which supplied the Baths of Caracalla. In *Leptiminus*, the strontium analyses applied to the sinter deposits of domestic cisterns indicate that they were partially filled with aqueduct water, not only with rain water (Bagust, Fletcher and Morgan 2011, 459, Yegül 1991, 393). This is comparable to later Islamic practices in Spain, by which the rural irrigation systems were diverted once a week to the city conduit in order to fill the private domestic cisterns (Glick and Kirchner 2000, 305). Lastly, the large cisterns of Constantinople (Crow, Bardill and Bayliss 2008, 15), built at least 30 years after the construction of the aqueduct, could, in fact, be related to the need for a larger water storage capacity prior to the expansion of the aqueduct to sources further away.

Of course, cisterns on their own (unless they have been analysed for strontium isotopes as in *Leptiminus*) cannot demonstrate the functioning of an aqueduct, so they have to be considered in their own context. In Tarragona, for instance, cisterns only appear in the upper town, next to the terminal tank of the aqueduct and far from the water table, whereas in the lower city, wells (and not cisterns) are more numerous.

Late Roman water legislation

Roman law included various provisions on aqueducts, regulations on who was responsible for their maintenance, and who actually owned the infrastructure (Brunn 2002a, 2012). The transition into the post-Roman period of the maintenance systems and the ownership of the aqueducts may shed some light on the evolution and abandonment of water supply systems.

Roman water law is based on the private (individual) ownership of springs and small rivers, and the communal nature of big (navigable) rivers, although water could be diverted from them if a permit was issued.[15] This means that aqueducts were, in fact, private

[15] Springs: *Dig.* XLIII 20.3.3 (*aqua, quae in rivo nascitur, tacite lucri fit ab eo qui dicitur*); Rivers: *Dig.* XLIII 12.2 (*quominus ex publico flumine ducatur aqua, nihil impedit nisi imperator aut senatus vetet*) and XXXIX 3.10.2.

property and responsibility of the city councils ('public' as opposed to 'communal'), who had the rights over the springs and rivers.

Whereas Frontinus gives a detailed account of how the water system of Rome worked in the early second century CE, comparable to what we know from other, provincial sources like the *Lex Ursonensis* and the *Lex Irnitana* (González Román 2011, 59), it is necessary to look at the Theodosian and Justinianic law codes to see how aqueduct legislation worked in Late Antiquity. The end of the early Roman municipal magistracies (but not of the curiales themselves; Curchin 2014, Fernández 2017, 165-70) blurs our view of late Roman control over municipal water resources, but it is evident that they remained under public control. However, in Late Antiquity the lack of public funds changed this situation, and landowners whose lands were crossed by a public aqueduct were freed from certain taxes, but had to be responsible for the cleaning and maintenance of those conduits:

> *Possessores, per quorum fines formarum meatus transeunt ab extraordinariis oneribus volemus esse immunes, ut eorum opera aquarum ductus sordibus oppleti mundentur nec ad aliud superindictae rei onus isdem possessoribus attinendis, ne circa res alias occupati repurgim formarum facere non occurrant.* (*CTheod* XV.2.1=*CIust* XI.43.1.1)

> 'We wish landowners whose lands are crossed by the course of aqueducts (*formarum*) to be exempt from extraordinary burdens (*oneribus extraordinariis*) so instead they may keep the aqueducts free from filth (*sordibus*), nor are there other requirements to be made from such owners, lest being occupied with other things they fail to clean the aqueducts'.

This transfer of responsibilities contrasts with the one provincial example of a municipal aqueduct expert, in fifth-century Vienne (*DLH* II.33). All this evidence gives an impression of a very fragmented system, where aqueduct control really varied from case to case, although until the sixth century all the evidence suggests varying degrees of continuous municipal ownership.

Comparanda

The fragmentation of the Roman world during Late Antiquity is evident not only in the political sphere, but also in the economic and social ones, as Chris Wickham has explained in his *magnum opus*, *Framing the Early Middle Ages* (2005) and this affected the way towns and urbanism developed in each different region. Thus, the way aqueducts fared in each region varied greatly, even within the Iberian Peninsula. Taking this into account, the use of comparanda from other regions of the Mediterranean will not be a precise way to analyse Spanish aqueducts. However, an overall picture of continuity or abandonment may serve as a good basis from which to develop arguments and theories about the Iberian aqueducts.

Firstly, it is clear that in those areas where strong local elites (especially bishops in later centuries) were present, or where state intervention is known, aqueducts seem to have continued in use during Late Antiquity: the elites and the state have the resources to invest in and the manpower to carry out these large works. Secondly, there is a visible gap between the seventh century, when it seems that the last Roman aqueducts ceased to function, and the eleventh, when old abandoned structures are put back in use.

Italy is, most certainly, the best-studied area. Not only the volume of the archaeological material favours Italy; the amount of available written evidence is also more abundant than that of any other region. Most of the aqueducts of the individual cities have been studied, although the later phases have been the subject of various specific works mostly by Paolo Squatriti (Squatriti 1998, Magnusson and Squatriti 2000) based on the earlier work of Bryan Ward-Perkins (1984) and, more recently, by Yuri Marano (2015). Rome has, as usual, drawn most of the attention both in ancient texts and modern studies (Coates-Stephens 1998, 2003a, 2003b, 2004), but Ravenna, the other main late antique centre has been taken into close consideration. In both cases it is not only aqueducts, but also baths and sewers that are restored, repaired and built. In Italy both the state and the Church (not only the Pope in Rome, but also other bishops like those of Naples and Nola) seem to have been actively involved in water-related constructions, although it is evident from the seventh century onwards that the general situation is one of decline, even in

Rome itself, and new urban water systems would not appear until the thirteenth century (Magnusson and Squatriti 2000, 246-50).

The most obvious point of comparison for Spain would be, however, South Gaul. Even if it is home to some of the most impressive Roman aqueducts, including the Pont du Gard, little is known in general about the late phases, and there is no general work that addresses this issue.[16] Most of the cities of Southern Gaul, like Lyons, Nîmes, Arles, Toulouse, Bordeaux or Marseilles seem to have preserved their aqueduct supply during the fifth and into the sixth centuries. This was the result, as in Italy, of direct state involvement during the last century of Roman rule (as in Narbonne – *CIL* XII 4355)[17] and of episcopal intervention. Besides, in Southern Gaul there are later medieval examples of hydraulic architecture, especially in the ninth century, like the conduit of the *fons Sancti Remigii*, although it seems to be largely unrelated to the Roman methods of aqueduct construction, for no levelling of the slope is noticeable, and it seems to be a crude ashlar conduit built in order to divert a stream (Wyss 1996, 188-9, 303). Later in the Carolingian period, just as in Italy, bishops can be linked to the construction of water-supply systems reusing the old abandoned Roman conduits, such as those of Beziers or that of Le Mans, repaired by Bishop Aldric (Benoit and Rouillard 2000). North Africa, in particular Morocco (*Tingitania*) is also a good point of comparison, which is in fact discussed at length in the appendix.

Lastly, the Eastern Mediterranean offers more examples of aqueduct continuity, and new late antique aqueduct and cistern construction, most famously in Constantinople (Crow, Bardill and

---

[16] I necessarily need to refer to various different publications (Bedon 1997, Bouet 2003, Fabre, Fiches and Paillet 1991a, 86-7), which summarise our current knowledge.

[17] *Pontem portas aqueduct(um) quaru(m) r[erum] usus longa incuria et vetustate co[...]rat civitati restauravit ac redd[idit] et ad Praeturianam Gall(iae) Pr(a)efect[uram] iudicio a(u)gust(a)e remuneration[nis causa]* = '...] restored and repaired the bridge, the gates and the aqueduct, whose use due to long neglect and old age [...], for the city [of Narbonne] and for the Praetorian Prefecture of Gaul for the sake of a venerable reward.'

Bayliss 2008). The Eastern Empire preserved its earlier functional and symbolic interest in aqueducts (Pickett 2017), which may explain the long-lasting continuity of hydraulic engineering absent in the West (Martínez Jiménez and González Gutiérrez 2017). There we also find numerous examples of episcopal water-related constructions, many attested through inscriptions (DiSegni 2002), and works such as those of Choricius of Gaza (Mayerson 1986). The Levant, despite the extended continuity of Roman domain offers more points of comparison with the Iberian Peninsula because of the overlap between Roman and Umayyad phases and construction techniques (Porath 2002), who showed in the East a similar contempt towards Roman aqueducts as in the West.

URBAN WATER SUPPLY: AQUEDUCTS AND MORE BEYOND THE ROMAN PERIOD

As explained above, the fifth century, with the collapse of the Roman administration and the emergence of local powers and Germanic kingdoms, clearly marked a point of no return for towns in Spain, deeply affecting their urbanism and infrastructure – and their aqueducts. Despite this, aqueducts were not the only source of water in urban sites, neither in Roman or post-Roman contexts. In this chapter we will look at the various urban water supply systems, paying particular attention to the fate of the aqueducts of the Iberian Peninsula, distinguishing between those which continued into the post 400 period and those that did not.

QUANTIFYING AQUEDUCTS IN THE POST-ROMAN PERIOD

In Roman *Hispania* a grand total of 69 urban aqueducts were built, unevenly distributed amongst 41 cities.[1] The ratio for *Hispania* roughly coincides with the data published for the 49 urban aqueducts of *Byzacena*, although over a much larger territory (Casagrande

[1] These include 65 known Roman aqueducts (Sánchez López and Martínez Jiménez 2016, 270), plus the new conduits discovered and brought to our attention since that publication, including one certain conduit from Tarazona (Bonilla Santander and Serrano Arnáez 2012, J. F. Casabona Sebastián 2014), one in *Caraca* (Gamo Pazos, Fernández Ortea and Sánchez Velasco 2017), the unpublished aqueduct of *Valeria* (López, Domínguez-Solera and Muñoz 2014), and the possible conduit of *Salionca* (Martínez Santa-Olalla 1931-1932).

2011). The majority of these aqueducts were built in secondary towns, and most were built during the last decades of the first century CE (Martínez Jiménez 2019). Their chronological evolution into Late Antiquity is marked by a steep decline during the third and fourth centuries, with this trend continuing into the fifth. During this period, most secondary nuclei (which had acquired aqueducts in the early Roman period) ceased to function as central places and saw their aqueducts going out of use (Houten 2018). After the late fifth and the sixth centuries, a small but consistent number of aqueducts remained in use into the Middle Ages (figure 6).

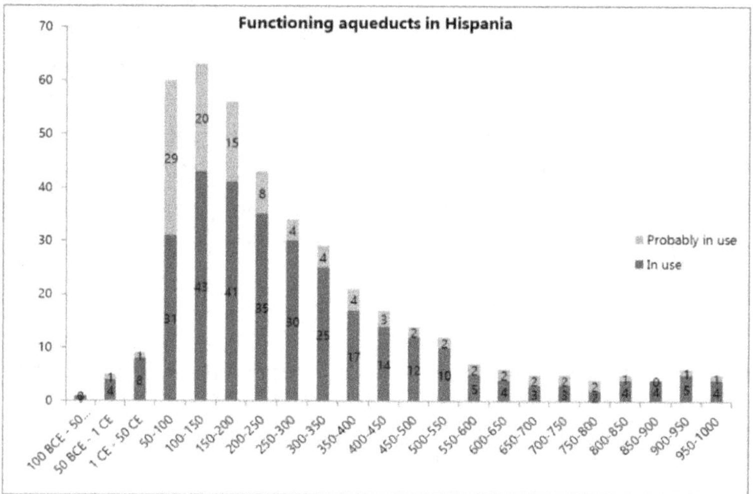

Figure 6: Graph depicting the number of functional aqueducts in the Iberian Peninsula (modified from the original in Sánchez López and Martínez Jiménez 2016)

Aqueducts dated to 400 CE

Before addressing those aqueducts which did continue into Late Antiquity, it is necessary, however, to focus on those which have been left out of this study, including both those that can definitely be dated as abandoned by the end of the fourth century and those for which there is not enough or conclusive evidence.

There are several Roman aqueducts where abandonment cannot be dated with certainty, due to the lack of available material, either from the aqueduct itself, or from the city that was supplied by it, many times because these are places which have been continuously occupied. The only datable evidence we have for these is that they are Roman, and that they were probably built during the first or second centuries CE. There are various aqueducts which we know only from inscriptions and for which we have no remains at all, like those of Cáparra, *Castulo* or *Egabrum*.

In some other cases we have partial evidence, either from the aqueduct or from the city itself that suggest an abandonment date during the fourth century or earlier. Amongst these we find several remarkable examples.

One of them is the aqueduct of Cádiz, the longest in the Peninsula (almost 75km), which crossed the bay into the island-city by means of several inverted siphons made of stone pipes that ran on the bottom of the lagoon, and the remains of which impressed even Islamic chroniclers (al-Makkari, I.6). Its course is roughly known thanks to recent surveys and excavations (Lara Medina 2018, Pérez Marrero 2011), and the evidence for hydraulic infrastructure within the city (sewers, cisterns) seems to suggest a second century abandonment date. Besides that, the whole city seems to have been greatly affected by events in the course of the third and fourth centuries, including the silting of the harbour, which had been in use since Phoenician times, and already the fourth century poet Avienus (267-8) mentions the city as: *nunc egena, nunc brevis, nunc destituta, nunc ruinarum ager est* = '[Cádiz] is now poor, now small, now abandoned, now a field of ruins'. The loss of importance and of rank, taking Avienus with caution, and the length and complexity of the aqueduct make us wonder about its continuity into Late Antiquity (Bernal Casasola 2018).

Another striking example is that of *Italica*, the first Roman colony in Spain. Even if much is known about the early town, the later phases are very obscure. The original foundation is currently under an early modern village, Santiponce, so very little is known of it. The large Hadrianic expansion of the city, however, is better known. During the fourth century most of this new urban area was not inhabited, and at some point in Late Antiquity the inhabitants built a new, smaller defensive wall that left outside most of the Hadrianic *urbs nova* (Rodríguez Hidalgo 1997). It is generally assumed that the area outside these new walls was largely abandoned, and that the population shifted to the fortified area. These were later restored by Liuvigild (or perhaps built then) during Hermengild's rebellion (Bicl., 584.1).[2] The city had become a bishopric at some point earlier in the fifth century, and a very active sixth-century mint (Caballos Rufino, Rodríguez Hidalgo and Marín Fatuarte 1999). The first-century aqueduct of *Italica* is relatively well known, and it is possible to distinguish two different phases: a first one related to the *urbs vetus* and a new expansion linked to the Hadrianic urban development (Canto 1978, Sánchez López and Martínez Jiménez 2016, 225-8). But as the aqueduct has not been excavated, and the baths were badly excavated in the early twentieth century, there is no direct dating evidence available from either (Caballos Rufino, Rodríguez Hidalgo and Marín Fatuarte 1999, García-Entero 2005, 709-17).

However, in most cases those without adequate dating evidence belong to secondary nuclei such as Consuegra, Chaves, *Uxama*, *Arucci*, Ronda, Jaén, *Ilipla*, *Egabrum*, *Lacipo*, *Ucubi*, Fuente Obejuna, or *Ocuri*, which appear to lose all urban functions in Late Antiquity.

Those aqueducts which, on the other hand, can be securely dated as abandoned by the end of the fourth century are fewer in number. Some of these can be linked to sites which ceased to exist by then, as for instance the sites of Los Bañales, *Begastri*, or *Caraca*. In some other cases, like *Ebusus*, Andelos or Tiermes, the terminal cisterns or the conduits of the aqueducts appear abandoned with plenty of fourth-century material retrieved from excavations, while the town preserves some degree of urban nature. In a couple of cases, we know of aqueducts damaged and abandoned as a consequence of earthquakes and never repaired, as the three aqueducts of *Baelo Claudia* or the *Aqua Augusta* of Córdoba. Lastly some single-purpose aqueducts linked to more or less still vibrant settlements in Late Antiquity fall into this category, such as that of León, built in order to supply the baths of the legionary camp, and that must have

[2] The chronology for these walls is undetermined beyond their late antique adscription (Ruiz Prieto 2013).

been in use only during the lifetime of these baths (in this case, up to the third century).

In all these cases, it appears that political importance and hierarchical centrality are essential for the continuity of urban life and the Roman water supply systems linked to it. And yet, we find two main late antique urban centres which had their aqueduct supply interrupted by the fourth century. These are Cartagena and Toledo. The truth is that both settlements, even if they had been important in the Republican period, had become secondary nuclei by the third century, which may account for the end of their aqueduct supplies, as the local elites perhaps could not maintain the water infrastructure. Only in the late sixth century did they become increasingly important, as Cartagena was turned into the capital of the Byzantine province of *Spania* and Toledo became the Visigothic royal capital. In the case of Cartagena, the aqueduct was already in disuse by the third century, as the materials retrieved from the distribution cistern and the Cerro del Molinillo indicate (Egea Vivancos 2002). Something similar can be said about Toledo, where the abandonment layers of the *castellum aquae* (the so-called 'caves of Hercules') are consistently fourth-century in date (Aranda Gutiérrez, Carrobles Santos and Isabel Sánchez 1997, 334-6). As will be discussed later, the new late antique relevance of these two sites did not lead to the revival of the old aqueduct systems; these were left abandoned and unused, even in the Islamic period.

Aqueducts continuing into the post-Roman period

In total, those aqueducts in the Iberian Peninsula which we cannot confirm to have been in use after 400 CE sum a total of 52, which is a 75.4%. For *Tingitania* (discussed in the appendix) the total is seven out of eight (87.5%), although with a sample much smaller and a worse archaeological knowledge, which make the figures not statistically relevant. In *Hispania*, the remaining 24.6% (17 aqueducts in 13 cities) have provided with some degree of evidence for continuity into the fifth century and beyond. These are the aqueducts of Almuñécar, Barcelona, Braga, *Conimbriga*, Córdoba, Lisbon, Lugo, Mérida, Segovia, Seville, Tarragona, Valencia, and Zaragoza (figure 7). To this list we need to add the newly built late antique aqueducts of *Reccopolis* and the *Qanat Amir* of Córdoba, which have not been taken into account for the figures on continuity. While all of these case studies will be explored in detail in the catalogue, it is necessary at this point to briefly introduce them, fur further reference later in following chapters.

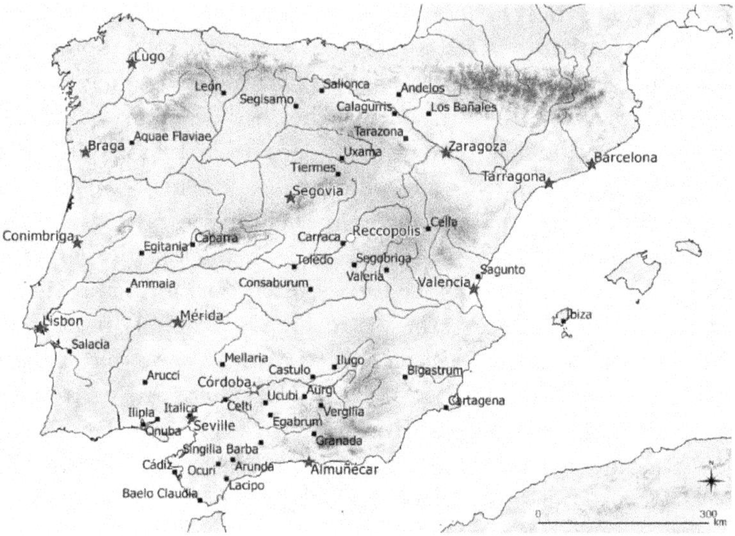

Figure 7: Map of the aqueducts discussed in the book, indicating those that were out of use in Late Antiquity, those for which we have evidence of continuity, and the newly built late antique aqueducts.

Almuñécar, a small coastal town in southern Spain originally founded by the Phoenicians, was a main trading harbour for the mountain inland areas of *Baetica* and also a main producer and export centre of *garum* and other fish products. Its aqueduct was probably built in the Flavian period, bringing water from some 7km away, and used to feed the city and the fish processing factories. While the city seems to have become largely abandoned by the time of the Byzantine invasion of the mid-sixth century, the factories seem to have been active into the late fifth. The abandonment date of these complexes is what suggests a possible late fifth century abandonment date for the aqueduct.

Barcelona was a Roman colony, which despite its secondary role in the early period, it became the main centre of north-eastern *Tarraconensis* during the Visigothic centuries, partly because of its strong fortifications, partly because of its powerful bishops. The latter were responsible for the construction of one of the best-known episcopal complexes of the Iberian Peninsula, although during the last decades, commercial archaeology has unearthed many other sites indicating an active urban centre in the sixth and seventh centuries. Its aqueduct tapped water from the mountains and flowed in two parallel conduits into the city, where various standing remains are still visible today. The chronology of the last pipe-fed baths and the abandonment layers from inside one of the conduits both point towards an abandonment date of the mid-seventh century.

Braga, the capital of the late province of *Gallaecia*, was an Augustan foundation. It became the capital of the kingdom of the Sueves, although they focused their construction efforts in the separate acropolis of Falperra rather than with monuments in the city. Most of the information about the city comes from the excavation at the *insula* 'das Carvalheiras', which has a bath complex abandoned between the late fourth and the early fifth centuries. The aqueduct tapped a series of small underground water mines, and while its construction must predate the second century, its abandonment can only be linked tentatively to that of the bath of the Carvalheiras dig.

Conimbriga was one of the first cities to build an aqueduct in the Iberian Peninsula, which is more remarkable considering that it was a native settlement and not a Roman foundation. The city was a main regional centre throughout the Roman period, but became a fortified frontier town in between the Suevic and the Visigothic kingdoms during the sixth century. There are various water consuming structures inside the late antique walls which were still in use between the mid fifth century and the alleged abandonment of the city, which are strong indicators for an end date for the aqueduct.

Córdoba was a Republican foundation, later expanded in the Augustan period, when it was promoted to capital of *Baetica*. Córdoba was a main city in the Visigothic period, and it became also the capital of the Umayyad state founded in the eighth century. Each of these phases has left its mark on the archaeology and monumentality of the city. In the Roman period, it was fed by four aqueducts: two leading into the city centre (and both out of use by the fourth century), one feeding the western suburb and another built to supply the tetrarchic palace complex of Cercadilla. The Cercadilla conduit seems to have been in use into the early eighth century, when it was put out of use Umayyad conquest of the city – as hinted by the sources and suggested by the archaeology of its *caput*. The Western aqueduct is one of the longest functioning aqueducts, as it is still in use today, even if during the tenth century it was diverted away from its Roman terminal point by the caliph in order to feed the Great Mosque.

Lisbon was the main harbour town of the western coast of Iberia and with links to Mérida. The Roman city itself is roughly known, and it seems to have had a flourishing late Roman phase, although little is known about it afterwards. The aqueduct is known only in its upper course, some 10km away from the city, but the presence of *garum* factories, public baths, and a new fountain, all in use in the fifth century, suggest a late fifth (if not early sixth) abandonment date.

Lugo was another Augustan foundation in northern Spain, established as a central place from where to control the local disperse populations and linked to the state mines. It was heavily fortified in Late Antiquity, but these reinforced walls preserved and respected the aqueduct, which had been built in the Flavian period. Even if the conduit has been excavated, these digs have not provided with any datable material for the abandonment. The archaeology of post-Roman Lugo is the great unknown, but it is clear that in the late fourth century Lugo had a strong elite and functioning baths, which may suggest that by the fifth century, when the city was taken by the Sueves, this was still the case.

Mérida was the religious and political capital of the Diocese of Spain, which included *Tingitania*, and it preserved its political importance during the Visigothic period and it showed this long-lasting power in the Umayyad period in the form of constant rebellions. Mérida has plenty of standing monuments and many excavated sites which provide us with a clear general view of the evolution of its urbanism from its Augustan foundation into the tenth century. It was originally supplied by four aqueducts, an early, foundational one (Las Abadías conduit) abandoned soon after, while the main three (Proserpina, Cornalvo and San Lázaro aqueducts) seem to have been in use into the fifth century according to their excavations. One of them might have been working into the sixth. This is further confirmed by the lack of water-consuming structures inside the walls and the collapse of sewers during the later Visigothic period. Furthermore, in Mérida we also have the remains of what seem to be a failed attempt to rebuild and repair one of the pillars of the aqueduct during the late sixth century.

Reccopolis was a new urban foundation of King Liuvigild in 578. The city, which represented the ideal of a late antique city, was built with walls, a basilica, suburbs, a palace, and also with an aqueduct, the first built in Iberia in many centuries, and the last one that would be built in the Roman technique. The course is roughly identified, but the lack of known water-consuming structures inside the city prevent from any hint towards an abandonment date, although it must predate the overall abandonment of the site in the mid-ninth century.

Segovia has the most iconic aqueduct in Spain, with its lofty arches of mortarless granite ashlars, and it was only put out of use in 1972. Even if very little is known about the city itself before the twelfth century, the characteristics of the water, the reduced volume of the flow, and the structural stability of the construction point towards a long-lasting aqueduct. The presence of medieval repairs prevents us from confirming a non-interrupted use until our days, but it is perfectly possible that it was in use during Late Antiquity despite its secondary rank within the Visigothic kingdom.

Seville was in Antiquity an ocean and river harbour, which can explain why it soon substituted *Italica* as the main city in western *Baetica*. Only recent excavations have unearthed late antique remains in context, like at the Patio de Banderas and the Plaza de la Encarnación digs. More importantly, various large structures connected to the water supply, including baths, sewers, and a terminal cistern have been excavated, and all their abandonment dates point towards an early sixth century date for the end of the aqueduct. The conduit itself is still standing in a few sections across the city, although in the nineteenth century over 400 arches were still visible. There is some controversy related to its current state, as it may reflect a later, Islamic phase of the aqueduct, but the abandonment dates, as given by the structures excavated inside the city, are unequivocal.

Tarragona, the city of the Scipios, was the first and the last city held by the Romans in the Iberian Peninsula. Built on a high rocky dome overlooking the coast by the main road to Italy, it was greatly expanded and heavily monumentalised by Augustus. This phase included two aqueducts tapping two different rivers (the Francolí aqueduct and the Gayá aqueduct), with a possible third (the suburb aqueduct) and an underground water mine linked directly to a monumental fountain (known as the *cuniculus* or 'tunnel'). The aqueduct feeding the suburb seem to have stopped running in the late Roman period, but the upper town conduits might have been functioning into the sixth. The complete redistribution of the settled area (shifting to two poles at the upper town and the harbour district) could be partially linked to a change in water accessibility. Added to this, there is a change in the functioning of various baths, an increasing number of cisterns (which could be linked to an aqueduct supply), and a possible late-fifth century fountain dedication, which support this chronology.

Valencia, located at the mouth of the Turia, and half way through between Cartagena and Tarragona, was a main commercial node on the Mediterranean coast. During Late Antiquity its bishops were responsible for the development of a Christian complex on the old forum area, the Almoina excavations, where a *nymphaeum*, a monumental fountain linked to the aqueduct was in use into the seventh century. The rest of the conduit is only partially known, and because it was extensively repaired during the Islamic period it is difficult to say if these works simply repaired and maintained an aqueduct which had been flowing in Late Antiquity or if they removed the Visigothic abandonment layers.

Lastly, Zaragoza is the main city of the Ebro valley, a city located at the crossing of the roads going from the Pyrenees into the meseta and between the Mediterranean and the Bay of Biscay. It was also an important river harbour and it was heavily fortified. This importance was retained during the Visigothic and Umayyad periods, although our archaeological knowledge for these phases is quite limited. Considering this, the aqueduct, which was certainly in use in the third century (but which was probably earlier), can only be said to have been working into the fifth century, when the last baths and drains were still in use.

WHY DID AQUEDUCTS CEASE TO FUNCTION?

No matter how, all Roman aqueducts at some point ceased to function. The immediate physical reasons behind this vary from active disruption to passive abandonment, but it may be possible to go a step further and look at the wider context searching for social, economic or political motivations.

Maintenance costs and expenses

There is nothing innovative or remarkable about the points which have just been made, as these are standard facts common to aqueducts across the Empire. A simple conclusion which can be drawn from them is that in Late Antiquity, for some reason, less care was taken in aqueduct maintenance and major repair.

We know from the sources that aqueducts were very expensive to build, but the maintenance they required in the long term also constituted a large expenditure (Martínez Jiménez 2019). It is difficult however to assess how much the maintenance of an aqueduct cost in Late Antiquity, although we know from several Gaulish examples that emperors and other high officials (such as the Prefect of Gaul) were responsible for the repairs of aqueducts in Metz (Halsall 1996, 239) and Narbonne (Solier 1991). Similarly, in Italy we have already mentioned that King Theoderic restored the aqueducts of Rome and Ravenna, amongst others. In the East, municipal authorities could not face large expenditure without seeking help from the central administration either (Saradi 2006, 345, Whittow 1990), and this certainly also included the maintenance of aqueducts, as we have seen for Caesarea Maritima.

If in the late period mayor repairs always appear related to munificence by the highest and richest members of the administration, it could be the case that these were too substantial to be covered by municipal funding (Curchin 2014). In the early Roman period we know of many various sources of municipal funding, including for aqueduct maintenance, specified in some cases as in the laws of *Dyrrachium* (Deniaux 2011) or *Irni* (González Román 2011). These expenses were probably covered by urban taxation, the *ex officio* payments of magistrates (*munera*), but also perhaps by means of the free labour offered by criminal convicts (suffering *damnatio ad opus publicum*) and public slaves (Duncan-Jones 1985, J. B. Ward-Perkins

1974, 14-7, 36-7). Some of this must have been preserved in Late Antiquity, as suggested by the mentions to *hydrophilakai* and the *servitudo aquarum* (*Var.*, III.31), although by the time of Constantine this may not have been enough. Later laws in the East (Saradi 2006: 344; cf. *Nov.* XVII.4, XXIV.3, XXV.4, XXVI.4 and XXX.8) specify that maintenance is responsibility of appointed office holders and not of civic councils may reflect this as well. Something which could have happened in the West too, where we have the *consulares aquarum* of Rome (*Var.*, V.38, VII.6), although other laws seem to suggest that the maintenance responsibility was passed on to those landowners whose lands were crossed by the aqueduct (*CTheod* XV.2.1 = *CIust* XI.43.1.1; Brunn 2000a, 2012).

Considering this, there is a pattern amongst those aqueducts which continued in use. Of the aqueducts which continued working (Almuñécar, Barcelona, *Conimbriga*, Córdoba, Lisbon, Lugo, Mérida, Segovia, Seville, Tarragona, Valencia, and Zaragoza), none seems to have undergone any major repairs after the fourth century – the aqueducts seem to have been kept in use through basic maintenance and simple cleaning duties. In late fifth-century Gaul Gregory of Tours (*DLH* II.33) mentions that in Vienne there was still a (public?) workman in charge of the maintenance of the aqueduct: *expulsus est inter citeros artifex ille cui de aqueducto cura manebat* = 'amongst those who were expelled [from Vienne, by the Burgundians] was the workman in charge of the aqueduct'. This could suggest similar people still existed in the Iberian Peninsula as well. This lack of major maintenance works and repairs, however suggests that, even if there was overall an intention to keep aqueducts functioning, when faced by serious structural problems there was little that could be done.

From what we know from construction and maintenance of public infrastructure in this period, there were skilled builders around. The work on the walls and the bridge of Mérida, the episcopal complexes of Barcelona and Valencia, the new suburb of Toledo, fortifications in Cartagena, and a long *et cetera,* all indicate this (Martínez Jiménez, Sastre de Diego and Tejerizo García 2018, ch. 5). Nevertheless, specialised builders do not necessarily mean specialised engineers. Perhaps it was not a lack of interest or a lack of funds that lay behind these abandonments, but rather simply a lack of skilled engineers capable of repairing and reconstructing these structures. The exceptional example of *Reccopolis* could indicate, nevertheless, that there was a demand for such skilled workers.

Figure 8: Sample sinter crust from the Valdepuentes aqueduct, Córdoba (Ventura Villanueva and Pizarro Berengena 2010: fig. 25).

The most usual damage caused to aqueducts is the formation of calcareous crystals: the so-called sinter, which, as already explained, forms on the interior of the *specus* in areas of hard water. Sinter crusts inside conduits cause damage in various ways; increasing the pressure, the speed of the flow, and reducing the quality of the water (Dessales 2011). The escalating speed and pressure is particularly dangerous for an over-ground conduit, because an increase in the pressure in a bend or corner could cause the aqueduct to tilt slowly outwards, causing its final collapse. In extreme cases, sinter could totally choke a conduit blocking the channel, forcing it to burst.

Sinter was frequently chiselled out of conduits, in order to keep them clean, as evidenced by the heaps of chiselled sinter found around some aqueducts, as in Nîmes (Fabre, Fiches and Paillet 1991b), It has also been suggested that boiling acids (like vinegar) may have been used to dissolve it inside pipes (Fahlbusch 1991, 7, 9-11). This must have been done on a regular basis, but at some point during the late Roman period it became a rare practice. For example, in Córdoba a sinter crust of over 12cm formed in the *specus* of the aqueduct of Valdepuentes (*Aqua Augusta*) during the third century (figure 8) (Ventura Villanueva and Pizarro Berengena 2010, 193-9). This is chronologically parallel to what happens at the aqueduct of Nîmes, although in the Spanish example the aqueduct was finally abandoned, whereas the French one was (badly) repaired and put back to use with a new lining of *opus signinum* on top of the sinter crust.

Other problems derived from a lack of proper maintenance include leaks and fissures, plant growths, and water pollution caused by an inadequate sediment decantation, mentioned in late Roman laws as *sordes*, which may seem minor complications, but which could develop into major problems (Dessales 2011). Seemingly, these kinds of minor problems were left unrepaired or unattended in minor towns in the third and fourth centuries, which led to the final abandonment of several aqueducts. This could be behind the abandonment of the *cuniculus* of Tarragona (Remolà Vallverdú and Ruiz de Arbulo 2002, 41-63), the aqueduct of Toledo (Aranda Gutiérrez, Carrobles Santos and Isabel Sánchez 1997), and that of Tiermes (Argente Oliver and Díaz Díaz 1980) amongst others.

In the case of sewers, it seems that the increasing accumulation of sediment (sometimes the result of conscious resurfacing of roads) made access to the drains increasingly difficult. These later phases also tended to block the gutters that flowed into the main sewers (Ruiz Bueno 2018a).

Vis maior and unavoidable damage

Of course, there were other events, unpredictable in nature, which could render aqueducts useless. Earthquakes were a major cause of damage to aqueducts, which are very susceptible to them. This is not only on the sections of the conduit carried on arches,[3] but also ground-level *substructiones* could be damaged by shifting terrain. This is especially true for the southern part of the Iberian Peninsula due to its high seismological activity, as the collapse of the aqueducts of Valdepuentes in Córdoba and the aqueducts of *Baelo Claudia* reflect. In fact, Cartagena is in a very seismically active area, and it is possible that earthquakes were the cause of the collapse of its aqueduct. Related to earthquakes, but not necessarily, is the alteration of the water table at the *capita aquarum*. If an aqueduct was fed by a spring and this spring's level was modified by seismic activity, then the supply would be disrupted. If the spring began to flow at a lower level, then the whole aqueduct was rendered useless. This has been proposed for the case of Metz, although not for any Spanish example as yet (Lefevbre 1997).

War and sieges have been seen usually as a source of unavoidable damage to aqueducts, but a brief *caveat* should be made. It is too easy to link the end of aqueducts to violent destruction during the course of a war or invasion, especially in this period in which Spain was sequentially invaded by the Germanic peoples, the Byzantines, the Franks and the Arabs, and was also deeply affected by civil wars. When a city is under siege, the besiegers intend to take the town, so radically damaging its aqueduct would seem to be counterproductive, although it is true that cutting the supply was a way to accelerate the fall of a city. The clearest example from this period is Rome: in the course of the Gothic wars Totila marched towards Rome and besieged Belisarius, who would not surrender, and then Totila cut the aqueducts that fed the city, in an attempt to force the population to hand over the city. This stratagem, however, did not work, because Rome had several springs inside its walls, and cisterns were plentiful. Actually, the main concern of the Romans was that they could not go to the baths anymore, because there was no water

[3] Vertical uprights in general are very weak against side movements, such as earthquakes; Cf. the identification of earthquake damage to columns at the site of El Tolmo de Minateda (Rodríguez Pascua, et al. 2013).

to supply them, as stated above (*Bell. Goth.*, I.19.28 and I.20.5). The overall damage to the aqueducts will not have been extensive or permanent, but will have consisted of selective blockings or piercings that could be easily repaired, as it happened in the year 554, when the imperial authorities issued the Pragmatic Sanction which included a clause that covered the restoration the aqueducts of Rome after the Ostrogothic siege.

Besides, none of the sources which mention the sieges of cities refer to damages done to aqueducts. Contemporary texts mention the sieges of Barcelona (673 CE: *HWR* 11; 801 CE: *VHI* 13), Braga (Hydat., 167 [174]), *Conimbriga* (Hydat., 237 [241]), Córdoba (*Hist. Goth.*, 45), Lisbon (Hydat., 181 [188], 240 [246]), Lugo (Hydat., 199 [204], 201 [206]), Mérida (Hydat., 111 [119], 240 [246]), Seville (Bicl., 583.1, 584.3), and Zaragoza (*DLH* III.29) and in no case are aqueducts even mentioned – regardless of the degree of destruction of the city. It is only in the case of the 711 Umayyad siege of Córdoba that we find a conduit mentioned and being blocked, although the sources for this are highly problematic and much later (as will be discussed below)

Just as mentioned above with the minor maintenance problems, if by any chance in the early Roman period an aqueduct collapsed or suffered major damage as happened in Arles, Cahors or *Baelo Claudia* (Raffard, et al. 2000, 125-7, Rigal 1997), there was the will (and resources, manpower and skilled labour) to put it back into use again, but this does not seem to have happened in Spain during Late Antiquity. It is therefore not so much the source of damage as the failure to repair it that needs further discussion.

Specialised workmen (or the lack thereof)

The end of highly specialised craftsmen in the west during Late Antiquity is not a new issue (B. Ward-Perkins 2005). The end of many of these aqueduct case-studies can be linked to major damage caused by the lack of proper repairs, even if they derive from long-lasting minor problems. Disregarding the regular maintenance that Spanish aqueducts may have been going through in their last centuries of existence (perhaps through *artifices* as that from Vienne), most of these structures were over three centuries old by Late Antiquity, and the cumulative effect of age, decay and minor problems could readily have caused their end.

Despite our fragmented knowledge of most Hispanic aqueducts, there is a lack of evidence for the late antique repairs or reconstructions evident in examples from Gaul (Cahors, Narbonne) or Italy (Rome, Ravenna). This means that by the sixth century, when we find again new attempts to repair old aqueducts and construct new ones (at Mérida and *Reccopolis*), over a hundred years had passed from the last major repair, which could be linked to the total loss of the technological skills necessary to maintain or repair an aqueduct. In Mérida we have new buttresses and reinforcement arches added at the Proserpina (Los Milagros) aqueduct during the mid-Empire (figure 5). These works were large scale structural repairs, carried out to reinforce the aqueduct pillars. This Roman work contrasts greatly with the Visigothic (sixth-century?) repairs that have been identified on the same aqueduct, built in a completely different construction technique (reusing different-sized ashlars and bricks whenever gaps could not be filled), which, according to Miguel Alba (Alba Calzado 2004, 255, Alba Calzado and Mateos Cruz 2008, 268), corresponds to other Visigothic constructions of Mérida.

The Visigothic repairs consisted of the reconstruction of one of the pillars that had collapsed, but there is no evidence to suggest that the arches or the conduit were ever finished (figure 9). The aqueduct's failure can be inferred from the lack of water-consuming buildings datable to the sixth or seventh centuries inside the walls of Mérida. Furthermore, the reconstructed pillar, located close to the surviving section of Roman arches, is separated by a gap from the settling tank. In this gap there are no traces of pillars or arches, so it is possible that these were never built. As Miguel Alba points out, this failed attempt at restoration, that there was the will and intention to preserve and (hopefully) restore the aqueducts one day, but there was a lack of knowledge and skill, or perhaps resources to complete the job (Martínez Jiménez 2012, 35-6). Re-erecting the pillars may have been a great achievement, but we cannot confirm that a successful restoration of the conduit took place. Levelling and creating a conduit that would fit into the old Roman one without altering the gradient required a knowledge of engineering and highly specialised tools and levelling equipment which I doubt were available in Visigothic Mérida.

Figure 9: Possible Visigothic pillar of the aqueduct of Los Milagros, in Mérida. Note how it tries to imitate the Roman technique, alternating ashlars with bricks, but also how even if starting as a solid prism, it begins to rotate on its axis and tilt to one side.

Similarly, the re-lining with *opus signinum* of the *specus* of the aqueduct of Segovia or of the pool of the *nymphaeum* in Valencia in the late Roman period were feasible repairs which were successfully achieved, but this does not mean that their builders could have built an aqueduct – especially in the late sixth century. *Opus signinum* was a very common technique still in this late period, even if it was less refined than it had been in the Roman past, and despite its increasingly popular use as a luxury pavement in churches and palaces, it was still used to line hydraulic infrastructures. And still, mortar mixing is not the same as calculating loads or levelling a conduit.

At first sight the aqueduct of *Reccopolis*, which was built *ex novo*, might suggest the survival in Iberia of considerable technical expertise. The context, however, is unique – *Reccopolis*, as a royal foundation, could have access to a proper engineer (Martínez Jiménez 2015). The whole construction of the city of *Reccopolis* is a feat of Visigothic building enterprise, but the presence of the aqueduct raises the issue of who built and designed it. The construction techniques and the materials are clearly local in nature but, in the complete absence of similar works elsewhere in Iberia, the persons responsible for the design and construction are unlikely to have been local – they may rather have been Byzantine. After all, Liuvigild was building his city in the year of his *decennalia*, and at a time in which hostilities against the Byzantine province had ceased (Bicl. 570.2, 571.3; *Hist. Goth.* 49). This would mean that Roman engineers could have collaborated or even been sent to Liuvigild by the imperial administration.

Engineers and architects are known to have been sent at request to foreign rulers (including potential enemies) by the emperors. Even itinerant specialised masons are known for Merovingian Gaul (Lebecq 1996, 298-9), but the prestige associated to complex architecture, as preserved in Byzantium (Cuomo 2007, 131-45), made them a rare and prized set of skilled workmen. In the late sixth century, Bayan I, the Avar khagan (king), requested builders from Constantinople who would build him a palace and a bath (which they did), although he then forced them to build a bridge over the Danube

(John Eph., VI.24; Menand., XXV.1).[4] Similarly, and perhaps more closely related to the *Reccopolis* example, is the early-seventh century request of Pope Gregory to the imperial authorities for the *vir clarissimus* Augustus (presumably with his team of workers) to be appointed as manager of the aqueducts, as he was the most capable man available in Ravenna (Gregory, *Epist.*, IX.4). This perhaps indicates that in Rome able men who could manage water supply were rare, for the aqueducts were then almost collapsing:

> *Praeterea ante aliquantum temporis experientiae tuae praeceperamus ut apud filium eminentissimum nostrum praefectum [paraetorium] ageret, quatenus cura formarum committi Augusto viro clarissimo debuisset, pro eo quod omnino sollicitus atque strenuus est. (...) Nam sic despiciuntur atque negleguntur formae ipsae, ut, nisi maior sollicitudo fuerit intra paucum tempus omnino depereant.* (Gregory, *Epist.*, XII.6.71–81)
>
> Furthermore, some time ago we ordered your Experience to take action before our most eminent son the [Praetorian] Prefect, so that care of the aqueducts (*cura formarum*) might be entrusted to the *vir clarissimus* Augustus, since he is a man of real concern and energy. (...) For the aqueducts themselves are so despised and neglected that, unless great concern is shown over them, they may shortly be totally destroyed

A century earlier, Theoderic could still muster enough resources and skilled men to preserve and repair the aqueducts of Rome and Ravenna; but even then, as evidenced from one of Cassiodorus' letters, a water expert had to come from North Africa at one point.[5] These examples do not demonstrate that Byzantine engineers were sent to Liuvigild, but they show that these skilled builders moved around the Mediterranean with some ease within established diplomatic networks, especially towards those areas where they were difficult to find, not skilled enough, or completely absent. They making their way to *Reccopolis* would not have been that difficult or unheard of.

---

[4] Cf. footnotes 174, 296 and 316 in the Blockley edition of Menander.

[5] *Var.*, III.53, although this was not an aqueduct expert.

The plausible presence of foreign engineers in Spain would have only affected the construction of the *Reccopolis* aqueduct (unless they were also responsible for the failed attempts in Mérida), so it should be considered as an exception rather than as the norm. These highly skilled craftsmen, even if they were common and abundant in the provinces in the first century CE (*Nulla provincia est, quae non peritos et ingeniosos homines habeat* = 'There is no province which lacks skilled and clever men' Pliny, *Epist.*, X.40), were too specialised and too focused on their own trade to adapt and survive into the fifth and sixth centuries. This can be explained as a result of the end of the training of military engineers through the army and the lack of demand for large public buildings, which probably put apprenticeships through civilian training to an end especially after the gap in the demand between the mid-fifth and late-sixth centuries (Martínez Jiménez and González Gutiérrez 2017). This, in turn, had a negative effect on aqueducts, which very occasionally needed a thorough repair, because even if there were funds and the desire to carry these repairs out, there was nobody around who could fix them.[6]

It would appear that only after the arrival of the Umayyads in the West (who tapped directly into the East Roman traditions of construction) we see major engineering ventures. They appear in the late eitght and certainly in the ninth centuries. This is clearly the case in Islamic contexts, such as the *Qanat Amir* and the other Umayyad aqueducts built in Córdoba (see below), but also in the Christian north. In Oviedo, around the 840s, Ramiro I built a large vaulted structure, Santa María del Naranco (Martínez Jiménez, Sastre de Diego and Tejerizo García 2018, 288), which the *Chronica Rotensis* (24) records as the first building of its kind: both vaulted and built without timbers (*ex murice et marmore sine lignis opere forniceo ... edificavit*). At roughly the same chronology, Charlemagne was carrying out large engineering projects, including the vaulted and arched palatine chapel in Aachen and the Danube-Rhine attempted canal, the *fossa Carolina* (Werther, et al. 2017).

---

[6] Here my conclusion is directly opposite to that of Fernando Aranda et al. (2017, 171).

LATE ANTIQUE ALTERNATIVES

During Late Antiquity and the slow decline in the number of functional aqueducts, 'traditional' supply systems, like wells and cisterns, became commonplace again. Not that they had ever been out of use, but they became, in those cities which lost their aqueducts, more necessary than they had been.

Ground water: Springs and wells

Springs continued to act as main sources of drinking water. The one with a known late antique chronology is the harbour spring Tarragona which was in use into the seventh century, and was fed by the karst aquifer of the city (the *cuniculus*). Here, even if the access points to the *cuniculus* in the lower city itself had been blocked during the fourth century, the overflow still reached the fountain, although by the Visigothic period the basin had been modified, and during the seventh century the lack of maintenance can probably explain why it silted up completely and was abandoned (Burès, García and Macias Solé 1998). We know of other naturally occurring springs inside cities, like those in Mérida found at Calle Arquitas and Plaza de Pizarro (Feijoo Martínez 2006), or in Málaga (Peral Bejarano 1995, 121). Similarly, Córdoba had, according to the Islamic sources, many springs, at least one per neighbourhood (a total of 21, as per al-Makkari, III.1).

In Cartagena, which is built on a hilly peninsula surrounded by marshes (figure 10), the water table is very high, and water is easily accessible with wells, but this water is not good for drinking as it is very brackish. The inhabitants of Cartagena, therefore, had to rely on natural springs for their drinking water. There are three springs in Cartagena which provide enough water all year round. One of them is conveniently located inside the city walls, at the Cerro de San José; two others are located outside the walled enclosure, the Fuente de Cubas (the source of the aqueduct) and the Fuente Santa, located two and three kilometres away respectively. It is known from written evidence of the Islamic and the Christian periods (in fact, well into the Modern period) that people walked to these springs to get their drinking water (Ramallo Asensio, Murcia Muñoz and Vizcaíno Sánchez 2010, 217).

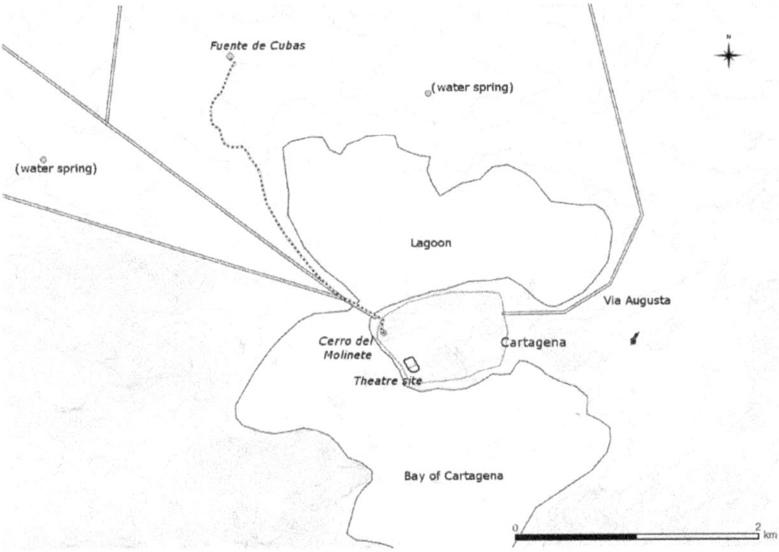

Figure 10: Map of Cartagena and its surroundings in Late Antiquity, highlighting the location of the springs and the course of the aqueduct.

In Málaga, a spring existed next to the theatre, which was monumentalised into a fountain, but once the theatre was abandoned, the water from this spring was piped to feed the *garum* factories (Peral Bejarano 1995, 121). It is conceivable that once the *garum* factories were abandoned (in this case, between the fourth and fifth centuries, certainly before the Byzantine invasion) the spring was again accessible for public use.

There is little evidence to suggest that these springs were diverted or used for any other purpose in Late Antiquity, but we have some examples of monumentalisation of springs, which can be seen perhaps as a continuation of the tradition of Roman *nymphaea*. One clear example of this could be the Foncalda spring at Oviedo, the ninth-century foundation capital of the Asturian kingdom, which was monumentalised with an ashlar aedicule and an *opus signinum* lined basin (figure 11) (Ríos González 1997). A similar example could be the spring near the church of Saint John in Baños, in Valladolid, which was originally built by King Recceswinth in 661, and around which a small aedicule was also built although this is not linked to any major settlement (Escorza 1997, Pérex Agorreta and Miró Alaix 2018, 285).[7]

Figure 11: Picture of the ninth-century monumental fountain of Foncalda, in Oviedo (photograph by Isaac Sastre).

Hot springs still remained as ideal sources for bathing complexes, and minero-medicinal waters preserved their importance into the early Middle Ages (Pérex Agorreta and Miró Alaix 2018). Lugo and Braga had baths with thermal springs, for instance, which had an extra aqueduct supply to bring cold water either to mix it with the hot water or to fill the *frigidaria* (García-Entero 2005, 292-3, López Quiroga 2004, 75, Morais 2011). However, once the aqueducts were out of use, and in order to keep the baths running, wells were dug, which enabled the baths to continue in use up to the fifth century. Similarly, at Orense, which was originally a small road-side *mansio*

[7] The church, as it stands now, is certainly not Visigothic as traditionally thought (Utrero Agudo 2006, 493), although it is built reusing materials from an earlier church.

that in the sixth century was promoted to a bishopric, the hot springs were still in use in this period (López Quiroga 2004, 87-90).

In other sites, baths not linked to aqueducts seem to have continued to function into the fifth and sixth centuries. For example, Pamplona, founded by Pompey the Great in Navarre, was an important town during late Roman times, and became increasingly important during the Visigothic period as the centre of the local nobility and an episcopal see. It became of key importance for the Visigothic kings in their campaigns against the Basques and, later, an Umayyad garrison seems also to have been established there (Faro Carballa, García-Barberena Unzu and Unzu Urmeneta 2008). Excavations in Pamplona have revealed that the town was largely supplied by wells, which were kept in use into the Medieval period. Its public baths, which have been excavated, also benefitted from the high water table, so into the fifth century the baths were fed by a well (figure 12) (Unzu Urmeneta, et al. 2006, 432-5). A similar situation can be described for *Complutum* (Alcalá de Henares), a vibrant and dynamic secondary town in the late Roman period. The various bathing complexes known all seem to have taken advantage of ground water for their functioning. In the case of the site of the so-called *domus* of Hypolitus, a medicinal well seems to have been used to feed a set of baths and a *nymphaeum* before the complex was built in the third or fourth century. The later building itself also had a bathing complex which reused the pre-existing structure. The large forum baths were similarly fed by underground water, although the extensive system of *specūs* could suggest a more complex system. A similar system existed at the new set of Visigothic baths of Mérida.

Large wells were also built in public areas linked to commercial activities in order to substitute the non-functioning, aqueduct-fed fountains. The two best examples of these are the fifth-century well of the late Roman market of Cartagena (Egea Vivancos 2002) and the 'great well' located in the forum of Valencia, built in the seventh century (Ribera Lacomba 2005, 228-33).

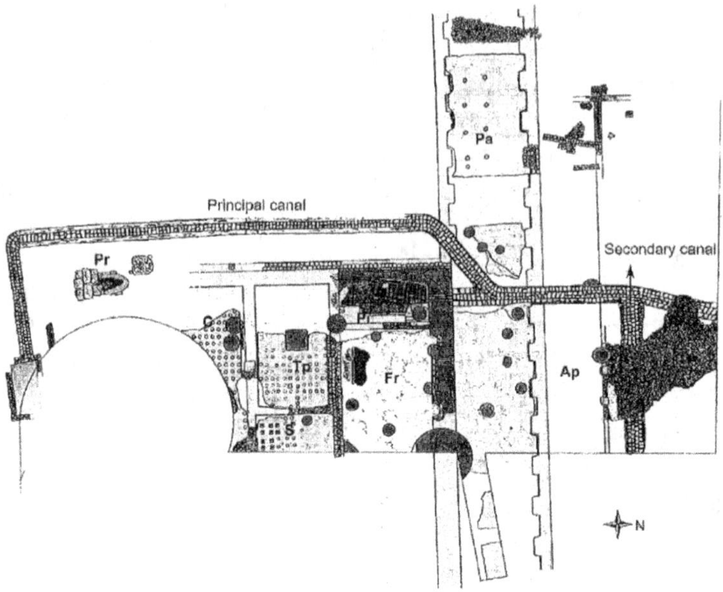

Figure 12: Plan of the Roman baths excavated in Pamplona (from Unzu Urmeneta, et al. 2006, fig. 1).

For domestic purposes, wells may have been the obvious ongoing solution. From the evidence available for the Islamic period of Málaga, it seems that the water table was high enough in the later medieval period (after the thirteenth century) so that neither cisterns nor conduits were necessary to supply the population (Peral Bejarano 1995, 123), which may reflect the nature of the earlier water supply (combined with springs). In Tarragona we know of an increasing number of wells in the harbour suburb, which is where the aqueducts had ceased to flow during Late Antiquity. At least two of them can be securely dated between the fifth and eighth centuries, and they are related to dwelling areas and not to workshops, which were the common structures of that area in the previous period (Macias Solé, Fiz Fernández, et al. 2007, #507). Another one was in use beyond the fifth century (Macias Solé, Fiz Fernández, et al. 2007, #403), and there are many others which cannot be securely dated. In Mérida the situation is very similar, and wells are present in the courtyards of many houses where the water table is easily reached (Gómez de Segura, et al. 2011, 130), as the excavations at the Calle Morería have shown. In this site, the old aristocratic *domus* was subdivided into several family dwellings. This reorganisation included the transformation of the old peristyle into a communal courtyard, in which a well was always built to supply the neighbours once the aqueducts had ceased to function (figure 13) (Alba Calzado 2005).

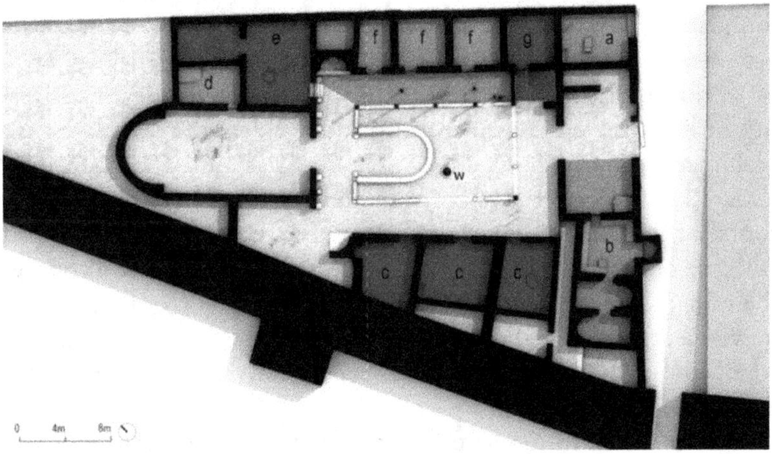

Figure 13: Late antique phase of the domus of the Marbles at the site of Morería, in Mérida. Note that the well in the central courtyard had become a focal communal element in the new layout of the house (Perich Roca and Gris Jeremías 2015, fig. 13).

Rainwater cisterns

Beyond these wells, there were other new water structures built in this period in cities which had lost their aqueducts. In Tarragona, for instance, the episcopal suburb of the Francolí river (built in the fifth century) was surrounded by some rural establishment with two new water supply systems (López Vilar 2006, 171-82). The first (in the excavation known as Sector VA) is a fifth-century cistern fed by a series of ceramic pipes, which probably collected rainwater from the rooftops. The second (in sector VB) is also a fifth-century large cistern, built not far away from the well; it is a rectangular structure (16 x 6m) built in *opus caementicium* with various buttresses, and lined in *opus signinum*. The presence of sinter may indicate that it was originally fed by the aqueduct (and not by rainwater) and, as it was an open-air structure, it has been suggested that it had a rural/irrigation function. The presence of a large well (1.3m in diameter and more than 4m deep) next to this structure has not been satisfactorily explained, because no water-lifting devices that could explain how the cistern could be fed from the well have been located (as has been suggested).

Cisterns also appear as the main source of water in new urban foundations. The site of El Tolmo de Minateda (near Hellín, in the province of Albacete), is a hill-top settlement, which has been excavated for the last 20 years by prof. Sonia Gutiérrez, from the University of Alicante, and has provided much evidence for the late Visigothic and early Islamic periods. The site was an old Roman *oppidum*, abandoned in the imperial period and then built again in the course of the late sixth century. The site has been correctly identified with the old Visigothic episcopal see of *Eio* or *Elo*, created to counter the imperial presence in the area, introducing a Visigothic centre of power very close to the frontier in a period (the reign of Reccared), when conflict was constant (Gutiérrez Lloret, Abad Casal and Gamo Parras 2005).

The late antique phase includes, as far as the excavations can tell, the re-erection of the city fortifications, and the construction of an episcopal complex (basilica, baptistery, and palace; figure 14), and of several houses. All of the constructions seem to have been built following a pre-conceived plan, and the architects chiselled the outline of the buildings on the bare rock, on top of which the buildings were erected (Abad Casal, et al. 2008, 329-30). The site itself is built on a rocky hill, and without a known aqueduct, so the new settlement had to rely on rainwater collected in cisterns, and the best known are those from the episcopal complex and those of the acropolis.

The acropolis is a fortified area on a spur of the hill, built on the opposite end from the entrance, and there two large rainwater cisterns have been identified. The other group of cisterns known are the pair located at the *episcopium*. These two were located on opposite sides of the basilica: north and south of the main building, and were fed by the rain water collected from the roofs, which was then led into the cisterns through rock-cut drains (one of which crosses the floor of the basilica; figure 15) and through natural cracks in the rock which may have acted as filters. These cisterns are relatively big (the north cistern measures 2.7 x 2.1 x 1.9m and the south one 3.9 x 2.5 x 2.5m), holding $10m^3$ and $24m^3$ respectively, and were lined in *opus signinum*. Calculations for the northern cistern show that it would have collected over $177m^3$ per year, so it could have been filled completely ten times every year, especially during the rainier months, perhaps enough to supply the clergy, but certainly not to be considered a viable public source of water (Martínez Jiménez 2017).

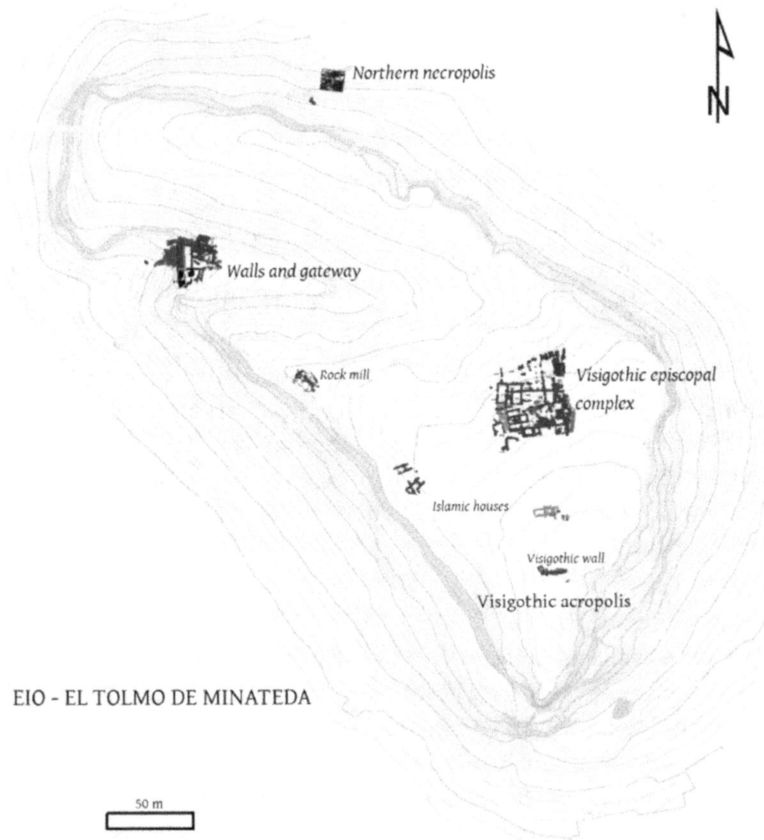

Figure 14: Plan of the excavations at El Tolmo de Minateda (Martínez Jiménez 2017, fig. 3).

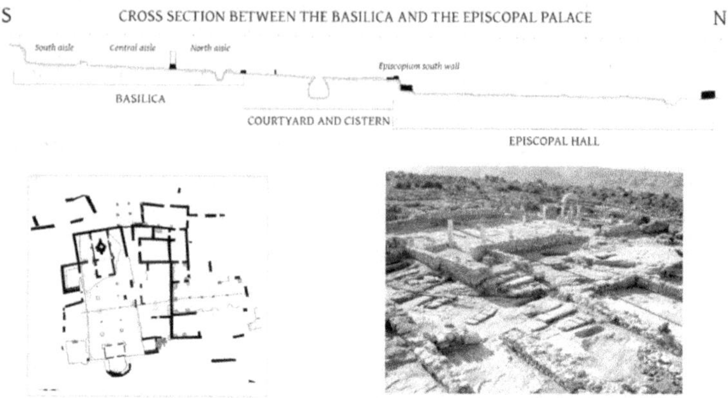

Figure 15: Cistern of El Tolmo (Martínez Jiménez 2017, fig. 6).

In the Islamic period, before the site was forcefully abandoned, the domestic structures that were built on top of the basilica and the *episcopium* do not have any cisterns or wells, although it is possible that the episcopal cisterns were still in use. Beyond this, there is evidence to suggest that liquid was stored inside houses in large ceramic containers (*tinajas*) which in rural Spain are traditionally used to hold water or wine (Gutiérrez Lloret and Cañavate Castejón 2010). It may be that these domestic water storage systems existed also in the pre-Islamic phases of *Eio* and elsewhere in the Peninsula.

***

From the available evidence, and with what can be obtained from parallel examples, it is clear that urban water supply became a priority for civic councils, and aqueducts played a very important role in providing water in quantity. This did not mean, however, that other ways of obtaining water, either private or public, ground or rain water, were dismissed. In fact, these alternative systems appear to come back as aqueducts disappear.

The end of aqueducts has to be related to the end of the administrative system that favoured their construction in the first place. Furthermore, involved and powerful urban elites have to be seen responsible for the preservation of aqueducts into Late Antiquity. This continuity was, nevertheless, limited in as much as the locally available municipal resources indicate that only basic maintenance was carried out. Any mayor construction works may have been too expensive at first and too technically challenging at later stages. In fact, trained engineers and specialised levelling tools were essential at this stage, but were most probably not available at all. In this context, the construction of the new aqueduct of *Reccopolis* should be seen as a *unicum* and not as a characteristic example.

Lastly, it should be kept in mind that aqueducts were part of a larger urban environment. Their continued presence made Roman town life possible in many cases, but they were not essential. The interaction between urbanism, aqueducts, and city life (and how they evolved through time) will be the main focus of the next chapter.

AQUEDUCTS, INFRASTRUCTURE AND URBANISM

Even if aqueducts were, in most cases, added at later stages in the development of cities, their construction caused in many cases a long-term transformation of the urban tissue and the inhabitants' way of life. Exceptions can include the aqueduct of Las Abadías in Mérida or the case of *Reccopolis*. By making life in the urban core easier and, generally, nicer, aqueducts became deeply integrated in city life. As a result, the slow decline affected towns in many ways.

THE PERSISTENCE OF A PUBLIC WATER SUPPLY

The public character of Roman aqueducts is best exemplified through street fountains, which made water readily available for many town dwellers, facilitating a denser population and a more comfortable urban life. While they are prominent in early Roman contexts, they are scarce in late antique ones. In fact, there is very limited evidence for the continued presence of public fountains fed by piped water in late antique Iberia.

In Mérida, for instance, we find that during Late Antiquity fountains were respected as far as they were still functional. During the late fourth and early fifth centuries there is evidence for their marble decorations being stripped out, but leaving the fountain otherwise intact – only when the water ceased to flow (or perhaps when it was evident that it would not flow again), were they fully dismantled (Alba Calzado 2007, 163, Mateos Cruz, et al. 2002, 77). A similar situation can be seen in Valencia. There, the pools of the forum *nymphaeum* were kept accessible throughout the Visigothic period, even if the fine marble fittings (which had been set up during a fifth-century redecoration) were taken down during the sixth century.

The pools and fountains, which were linked to the aqueduct, were left untouched (Albiach Descal, Ballester Martínez and Rosselló Mesquida 1999, 418-24). In both cases we see city dwellers acknowledging and appreciating water sources and, more probably, local authorities preventing any damage to functional public infrastructure.

Besides these examples of maintenance and continuity, there are also some examples for new fountains built in the late antique period, equally underlining the resilience of municipal concerns. The best example is the fountain identified at the site of Sommer, in Lisbon (Silva and De Man 2013, 397, Ribeiro, et al. 2017). This fountain is itself a small vaulted structure covering a pool built with reused material and lined with *opus signinum*, fed by a pipe, and located on a street corner – much in the style of a Pompeian *lacus* (figure 16), which appears to have been built during the fifth century and abandoned during the sixth century. Another (possible) late Roman fountain might have existed in Tarragona, as suggested by one single find. The inscription in question (figure 17), *CIL* II 4109, is a small plaque (40 x 45cm) with a small hole in its centre which is certainly contemporary with the inscription (since the letters avoid the hole), and it was found at Camino del Cementerio n. 7. As the inscription was discovered in the nineteenth century in the wall of a house, it is difficult to identify where it came from, but if it was originally found nearby, it would come from La Oliva hill, where the Gayá aqueducts passed. The inscription of the plaque reads: *BF.S DD(omini) NN(ostri) Leonis et Anthemi Augg(ustorum)*, which translates as 'BFS (?) of our lords emperors Leo and Anthemius', and has thus been dated to 468-473 CE (Pérez Martínez 2014). The letters BFS have no straight forward translation, although several interpretations have been proposed,[1] which include *B(onum) F(actum) S(aluti)*, 'A good thing, made for the salvation...' and *B(eatissimo) F(elicissimo) S(aeculo)*, 'In the most blessed and happy times...'. It would be tempting to see the F being the abbreviation of *fons* (for example *bonae fons saluti*, 'the fountain to the good health...'?), although other public fountains with inscriptions of Spain (like those in Córdoba;

[1] Cf. http://laststatues.classics.ox.ac.uk, LSA-81986 (Ch. Witschel)

CIL II^{2} 7.217-8) use *lacus* instead to refer to the fountain basin. If this inscription were linked to a fountain, the hole through the plaque may have been for the spout or the pipe, now gone. It measures roughly 10cm in diameter, which is the equivalent diametre to an *octonaria* (8-digit pipe; *Aq.*, I.42), which is just only slightly bigger in diameter than the standard pipe (the *quinaria* or 5-digit pipe), but contained just over twice the amount of water than the *quinaria*.

Figure 16: The late Roman fountain excavated at the Sommer site (Lisbon), currently displayed inside a hotel, consists of a large pool, fed by a pipe and covered by a concrete vault.

In the post-Roman period, especially in contexts directly connected to new political powers, we find the construction of new public fountains. In the Visigothic period we have *Reccopolis*, where a structure built by the side of the main road leading to the palace has many similarities with Roman public *laci*, like the abovementioned example from Lisbon. The large covered vat is considered to have been a public cistern (figure 18), which might have been fed by the water conduit identified at the palatial complex (Henning, et al. 2019, 742), perhaps a secondary branch from the conduit that fed the upper city.

Figure 17: CIL II 4109, inscription that may relate to a fountain head dated to the late fifth century, in Tarragona.

Furthermore, considering its position (at a lower level than the palace complex), it is indeed likely that the cistern was filled by aqueduct water, especially because it is not adjacent to any roofed surface big enough to have secured some water (figure 19) and it does not go deep enough to encounter the water table (Martínez Jiménez 2015). In the Islamic period, and especially in Córdoba, new fountains were built: one of them at the Mosque and another outside the *alcázar*. The *alcázar* fountain (the so-called 'fountain of the pipe') was fed by the new, eighth-century Alcázar conduit, and is mentioned as still existing in the tenth century, though we do not know when it was built. It seems to have been fed by the overflow (the unused) water of the *alcázar* itself as a sign of the piety of the caliph Abd al-Rahman III (Ventura Villanueva 2002, 126). The mosque fountains built in the 960s were (and are still) fed by the Umayyad diversion which tapped the water away from the Western aqueduct of Córdoba, rendering the former useless. Despite their obvious public nature, the mosque fountains have an essential ritual and sacred meaning, and it is doubtful that these were used as public sources of drinking or domestic water (like the *alcázar* fountain).

Figure 18: Public fountain basin/cistern of *Reccopolis*.

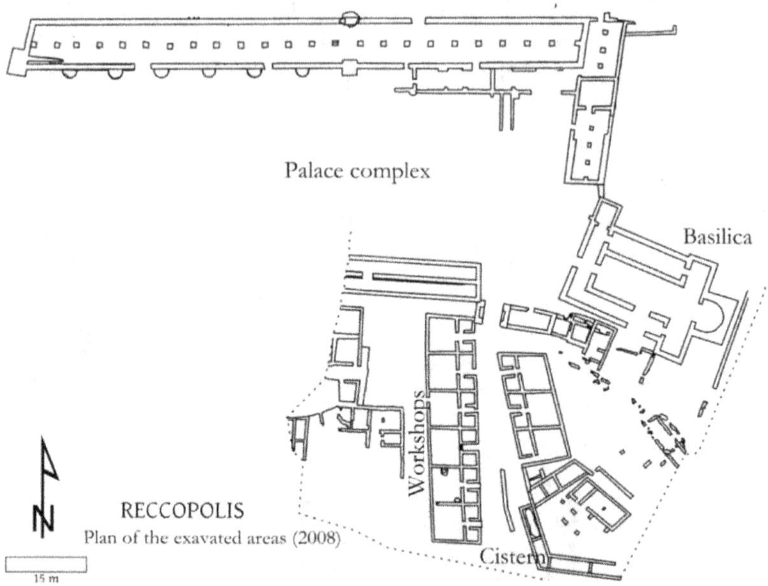

Figure 19: Plan of the excavations of *Reccopolis* (based on Olmo Enciso 2008, fig. 3).

The evidently public nature of aqueducts is exemplified in the maintenance and construction of fountains throughout late Antiquity. In the fifth century we still find municipal interest in preserving and constructing new fountains. By the end of the sixth century it is clear that the administrative involvement is limited to preserve what exists, even when the ornamental elements are taken down, which echoes similar contexts in Nîmes (Fabre, Fiches and Paillet 1991b) and Italy (*Var.*, II.7). As in many other aspects, *Reccopolis* appears as aunique example of post-Roman hydraulic munificence, with a possible public fountain. By the Islamic period, in the eighth and ninth centuries, this concept has been abandoned. The fountains that we know of do not form part of an integrated, long-distance, public supply system; they exist as either pious donations to a dynastic monument (which deprived a suburb from its Roman public water supply) or as run-offs of the private supply to the royal castle.

PRIVATE CONSUMPTION AND DOMESTIC ARCHITECTURE

The evidence for the continuity of private water supply is even scantier than that for public supply, and does not persist beyond the sixth century. In fact, only Mérida, Seville, and Barcelona offer documented excavated examples of private houses with hydraulic infrastructure.

On the one hand, we know of the continuity of various baths in urban *domūs* of Barcelona, including those of Bisbe Caçador and of Pati Llimona, in use into the sixth century, and seemingly part of large single-family houses (García-Entero 2005, 207-11). On the other, we have the houses of Mérida, like those excavated at the site of Morería, which in their fifth-century phase lacked piped water and were fed by a single well, and ended up being subdivided into several dwellings (Alba Calzado 2005). The *domus* of Cantaber, in *Conimbriga*, was a large urban *domus* supplied by pipes, and up to the very last moments of its life (early in the fifth century), the pipes were running. In fact, it is possible to tell that the house was abandoned because the hydraulic system failed and flooded the house at one point, never to be repaired (García-Entero 2005, 567). In Seville the houses at the Plaza de la Encarnación site had a lead pipe still in fifth-century contexts, although this is not evidence enough to suggest that Isidore's aforementioned knowledge of pipes as part of the house was first hand (González Acuña 2011, 380-1).

However, the very same elite *domus* of Barcelona which preserved their baths and their status during the sixth century, and that were building their own private sewers in the fifth, eventually went through a process parallel to those houses of Morería in Mérida. The sixth century phases of the *domūs* of Bisbe Caçador and San Honorat show not only an abandonment of the bathing structures, but also a subdivision of the house into multi-family dwellings (Beltrán de Heredia 2008, 282-3). This chronological coincidence implies a direct causal relationship between the existence of private luxury *domūs* and private baths. This is a wider phenomenon, as evidenced in the towns of the Byzantine province, which show no traces of elite urban housing whatsoever for the Byzantine period (Vizcaíno Sánchez 2009, 387-92). In fact, just as the houses of Morería in Mérida focus around the communal well, in the theatre of Cartagena, which had been turned into a dwelling area in the course of the sixth century, a communal well also appears (Vizcaíno Sánchez 2009, 400). There is no information about the *domūs* of Barcelona, because the houses could not be excavated in open area, but the hydraulic infrastructure of the baths may have turned into a communal fountain, parallel to Mérida or Cartagena, taking advantage of the pipe. Of Seville not much else is known.

Lastly, the end of public water supply to private individuals may also be behind the appearance of domestic cesspits, a development which is parallel to the decline of public sewer systems. Cesspits are known in dwelling areas in Cartagena (at the theatre) and in Mérida already in the fifth, but mostly during the sixth century. In *Reccopolis* cesspits appear in the Islamic period, although before that there does not seem to have been any proper covered sewer system (Sanz Paratcha 2008).

SETTLEMENT REORGANISATION

The availability of drinking water has always been one of the key factors which determined the creation of settlements. The presence of springs, or of a river and a high water table, was normally required from an area on which a settlement could be established, although this was not an imperative.[2] Aqueducts were normally built long after a city had been founded but, when built, enabled its further development and growth (Hodge 1989, 128-9) and, as mentioned earlier, aqueducts could actually remove the necessity of having a ready *in situ* water source (although this would be a risky strategy). This is especially true for Rome, which built its first aqueducts in the Republican period in order to bring drinking water for the population, though later aqueducts can be linked to the construction of large bath complexes or the *naumachia* (Hodge 2000a, 46-7, Stambaugh 1988, 130-1). Similarly, Constantinople needed its aqueducts because local sources could not sustain such a large population.

---

[2] Especially because the Romans' mastery over aqueduct construction enabled them to settle wherever they thought it suitable, without the imperative need of an immediate water source (J. B. Ward-Perkins 1974, 30-4).

But are the large cities of Rome and Constantinople fair comparisons? How much did small provincial towns, like the Iberian and Moroccan case studies, really need piped water? Recent studies on the evolution of settlement patterns in Crete indicate that in at least three cases (Polyrhennia, Cnossus, and the provincial capital, Gortyna) the main settlement shifted to a lower location once they got an aqueduct, indicating that piped water became essential for the new urban development (Kelly 2006). This strong link between aqueduct supply and urban settlement pattern can partly explain the different transformations that can be seen in the cities of late antique *Hispania* and *Tingitania*. These changes are two: concentration of settlement around those areas still supplied by aqueducts and the shift away from areas where the aqueduct has ceased to function.

Clustering settlements around aqueducts

The aqueducts, providing an abundant flow of water allowed settlements to preserve their original density, although in cases with a malfunctioning supply, this led to a concentration of population around the sources of water if other alternatives were not evident or accessible.

An example of the former case could be Barcelona, Seville or Lugo, where aqueducts working into the post-Roman period, together with a combination of accessible water table and a network of domestic cisterns, ensured that the settlement pattern remained largely unchanged. Eventually, however, it appears that the as the population dropped, the settled area of Barcelona slowly gravitated towards the episcopal complex, which was built at the entrance point of the aqueduct. In some other cases, however, the evident decline in population and subsequent re-configuration of the occupied area points towards aqueducts as elements attracting the population.

This is clear in the evolution of the Roman suburbs of Córdoba. Despite knowing through archaeology of the presence of suburbs (*vici*) in Córdoba, the Latin sources are silent on them, except for the one mentioned in a Roman inscription.[3] Only very late, in the Islamic period, do we see new ones mentioned: the suburb of *Šaqunda* (from its Latin local name, *Corduba Secunda*) was perhaps one of the most important, located south of the river, which is where the Muslim troops first settled (Casal García 2008). The old Roman western *vicus* was largely abandoned in the fourth century, and the Latin chronicles, in their brief mentions of this area of Córdoba, only talk about the church of Saint Acisclus, around which King Agila stationed his troops during the siege of 550 (*Hist. Goth.* 45), and where the last Visigothic soldiers held their ground during the siege of 711 (Arjona Castro 2001, 21-7). The Islamic sources also mention that this area was still inhabited as a Christian suburb into the ninth century (Sánchez Ramos 2010, 37). In this area we know the Western suburb aqueduct and the Cercadilla conduit; archaeology shows that the former was still in use in the tenth century, when it was diverted to feed the Mosque, and the latter went out of use in the early eighth. This contrasts with the decline of the eastern *vicus*, which was clearly out of use by the fourth century, perhaps because of the end of the Arroyo Pedroche aqueduct which supplied this area. Here no preexisting suburb is mentioned in the Arabic sources.[4]

In Tarragona this concentration is clearer. Tarragona, which had a very particular urban tripartite distribution (upper city, lower city and harbour suburb; figure 20) in the Classical period, shows a dramatic transformation in Late Antiquity (Macias Solé 2008, 296). The archaeological record shows population decline and the abandonment of large areas of the city during the Visigothic centuries. These changes were largely the result of underlying political and economic decline, but the details of how they affected the topography of Tarragona were probably influenced by the survival of one of its aqueducts. The upper town had been the centre of provincial administration and the focus of political life in Roman times, with a large

---

[3] *CIL* II[2] 7.274 mentions the *vicus forensis*, the 'vicus of the forum', although this seems to have been an intramuros neighbourhood.

[4] This is also reflected in excavations, as it was all turned into a necropolis (Sánchez Ramos 2010, 33-4).

forum and a temple complex for the imperial cult. This forum area seems to have kept its public function throughout the fifth century (Tarragona was the last Roman town to be lost to the Visigoths in 475 CE), but soon thereafter it was encroached upon, losing its public function. However, during the sixth century there is evidence to suggest that the episcopal complex was moved from the Francolí suburb to this area; the principal reasons for this move were probably because the new site was located inside the walled enclosure, because it was built into the old centre of Roman power (claiming continuity of authority), and because it was close to several large walls and other impressive structures (including large cisterns) that have been excavated close-by (Bosch Puche, et al. 2005).

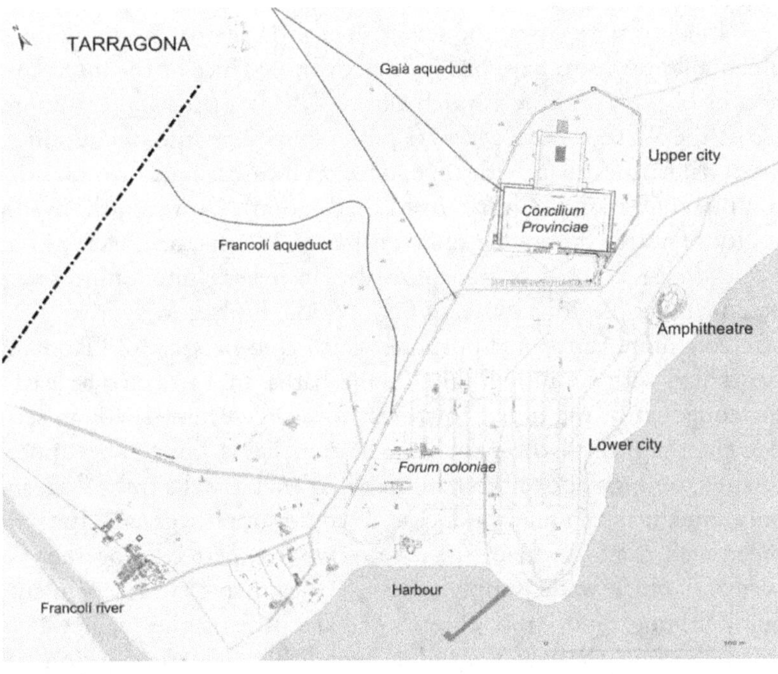

Figure 20: Plan of Tarragona in Late Antiquity, highlighting the course of the aqueducts and the tri-partite division of the old Roman city (based on Macias Solé, Fiz Fernández, et al. 2007, fase V).

A similar case can be made for Valencia and Braga. In Valencia the settled area appears to have concentrated around the episcopal complex, which was in the direct proximity of the functioning aqueduct supply: on the forum, adjacent to the *nymphaeum* and by the *castellum aquae* (Martínez Jiménez 2011, Ribera Lacomba 2008). In Braga, the correlation between settlement and aqueduct is less tenuous but, just as in Barcelona and Valencia, the cathedral complex was the attracting focus of the settlement, and the aqueduct terminal point is thought to be in the same area of the Carvalheiras site (Martins 2005, Martins, Magalhães, et al. 2016). Lastly, a comparable case can be put forward for the site of Casa Herrera outside Mérida, which even if it is not urban in nature, a settlement developed there in the sixth century, at a point where the water still flowed even if, at that time, it did not reach the city any more (Sastre de Diego and Martínez Jiménez 2013).

In all of these cases, the settled area and the post-Roman monumentality seems to have been a direct consequence of the intervention of bishops and the Church elites.[5] This is a situation common across the West, especially where new suburbs emerge surrounding martyrial shrines that, with time, develop into basilicas or monastic complexes (Esmonde Cleary 2013, Loseby 2006). However, the availability of water cannot be underestimated as a factor. Firstly, in a world where aqueducts were increasingly fewer, functioning ones became more precious sources of civic pride. But secondly, they also offered a more functional purpose, which is to preserve old Roman city living, with running fountains and baths. In Tarragona at least, the aqueduct in the upper town was to a large extent enabling settlement there until the network of cisterns was built. The upper town is too high above the water table to enable wells to be dug, so only aqueducts and rain water cisterns could supply this area. In fact, the aqueduct and some of the cisterns worked perfectly together as part of a single water supply system, rather than two independent ones, although inevitably cisterns ended up being the main water source beyond the late Visigothic period. The superimposition of cathedral complexes over Roman civic centres (directly connected to

[5] Lay, civil elites were probably involved too, but they are less obvious in the architecture of these areas.

water display and distribution) is clear in Tarragona and Valencia, while in Barcelona and Braga the episcopal constructions are located in the immediate vicinity of the *castellum aquae*, when not directly at it, as in the Western suburb of Córdoba.

A last example, which does not follow the previous patterns, but still shows population clustering around a functional aqueduct, should be mentioned. In Segovia, whose aqueduct seems to have been in use throughout the Middle Ages, suburban settlements appear to have emerged on the sides of the aqueduct where the *specus* could be (illegally) tapped. When describing the city in the thirteenth century, Archbishop Ximénez de Rada still mentions it as a high-ground settlement, indicating where the main settlement focus was, despite the already existing suburbs originated around the new Romanesque churches of Saint Martin, Saint Michael and Saint Andrew. These are all located along the lower reaches of the aqueduct from which they could tap water out of the conduit. This is still reflected in modern street names adjacent to the aqueduct: *Calle de [los] Cañuelos* ('Water-pipe Street') and *Calle de los Batanes* ('Fullers' Street'), which date at least from the late Medieval period.[6] This later medieval 'theft' of water can also be seen in some of the almunias (Umayyad rural estates, *al-muniya*) of Córdoba, the clearest example being that of Al Rusafa. The site (currently known as Arruzafa, but originally deriving from the Arabic for 'garden', رصافة) was built by Abd al-Rahman I in the eighth century, and in order to supply his estate he diverted the Cercadilla aqueduct at its source (Murillo Redondo 2009). In this case, however, the water-reuse did not develop into a new settlement.

Spreading for new sources

In most other cases we find that aqueducts ceased to function, and it is this lack of aqueduct supply which causes the shift of the settled areas. Those sites which had relied on aqueduct water to have a

---

[6] This practice of stealing water is recorded even in the 1860s (Fernández Casado 2008 [1972], 114).

densely settled urban area were forced to redistribute their population.

In Córdoba, for instance, while the suburb was drawn towards the aqueduct-fed area, the intramuros settlement, deprived of its piped water shifted in the opposite direction. Even if the first *oppidum* was built on the hilltop for defensive reasons, from the fifth century onwards, the population of Córdoba slowly began to shift away towards the area closer to the river, the theatre spring, and the water table. This area was also of political and strategic importance, as it is where the bridge, and ultimately, the Visigothic *palatium* (figure 21) were located.[7] The forum had already ceased to function as such during the fourth century and throughout the fifth it was encroached upon and dismantled (including its fountain) by small houses, and finally turned into a series of dumps and small agricultural patches of land (Márquez Moreno and Ventura Villanueva 2005, 438). This reconfiguration of the upper town during the fifth century ended in the sixth with an overall abandonment, parallel to the development in the south-western corner of the lower city of the episcopal complex and the Visigothic *palatium*. This could be further confirmed by intra-mural burials, which mostly appear in the northern half of the city (Sánchez Ramos 2010).[8]

A similar shift can be seen in Tarragona. While the walled city shifted its population towards the upper city, the lower city fed by the *cuniculus* (the water mine dug into the lower city's karst), and by the Gayá aqueduct, seems to have been mostly abandoned and turned into a semi-rural area (Burès, García and Macias Solé 1998, 79). The population appears to have also shifted to the harbour suburb. The suburb, which expanded from the mouth of the Francolí river and the Christian complex to the extra-mural harbour, became more densely inhabited in this period. In this area it is not only wells

[7] And, according to most authors (Marfil Ruíz 2000a), the Visigothic episcopal complex and cathedral. This location is, on a closer analysis of the archaeological evidence, doubtful (Arce Sainz 2015, Sánchez Velasco 2018, 94-5).

[8] Although some exist in the southern half as well (Ruiz Bueno 2016, 335, 435).

but also springs (both listed above) were numerous, favouring the relocation of the inhabitants of the lower town.

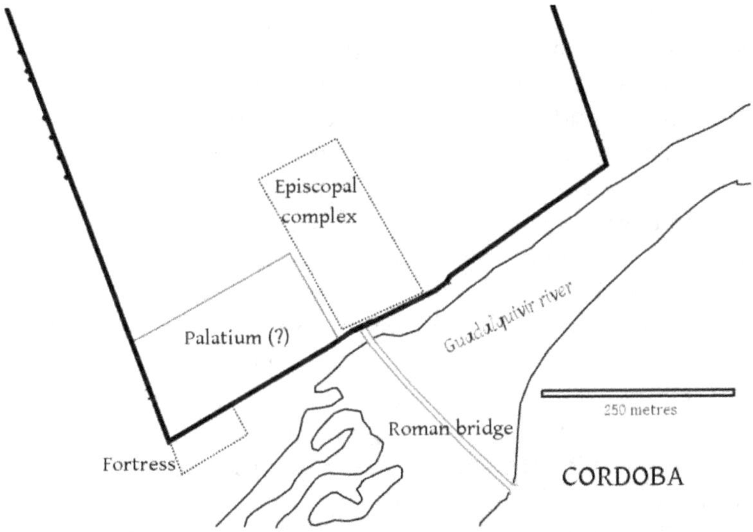

Figure 21: Reconstruction of the location of the palatium of Córdoba, and the supposed episcopal complex during the Visigothic period.

This situation, shift of the inhabited area because of the impossibility to dig wells, can also be seen in Toledo, built on a granitic dome on a bend of the Tagus. The city had two aqueducts, including one which brought water from quite a considerable distance (almost 75km) from a dam south of the river, and on reaching the city, crossed the Tagus on a siphon bridge, a considerable feat of engineering (Arenillas Parra and Barahona Oviedo 2008, Aranda Gutiérrez, Carrobles Santos and Isabel Sánchez 1997). It became the capital of the Visigothic kingdom during the mid-sixth century, but by this stage, the old aqueduct had ceased to function, as the fourth-century abandonment layers of the terminal cisterns, known as the 'caves of Hercules' (including the minor ones at Delegación de Hacienda and Calle Tornerías), and one of the upper town baths seem to indicate. This means that by this stage the upper city could only rely on rain water cisterns for its supply, although we know from the Islamic period onwards of the existence of the *assaqqá* ('azacán' in modern Spanish) – watermen who carried jars of water from the river up to the city on donkeys (Fernández Casado 2008 [1972], 210) - and of wells which did not provide drinking water.[9]

Having said this, the archaeology of the old upper city of Toledo is largely unknown, as it has been constantly inhabited ever since it was first settled, but the area immediately outside the city, the Vega Baja ('lower meadow') has not been touched by modern building. In fact, it has been a field ever since the ninth century, and despite being so close to the monumental centre of Toledo, the construction in the eighteenth century of the Royal Weapons Factory protected the area from any urban development until 2006. It was then that the emergency excavations prior to the construction of an important department store uncovered the remains of a large suburb dated to the Visigothic period (Olmo Enciso 2009, but with a big caveat–Moreno Martín 2018). This suburb, which not only has an organised street grid, but residential and workshop quarters as well, is over 70ha in extension, and located near the Roman circus. There, recent excavations have revealed that there were also some very important buildings built in ashlar blocks, floors paved in *opus signinum*, and other elements which have linked this site with the *praetorium suburbanum*, the Visigothic palace complex and its churches mentioned in the sources (Olmo Enciso 2009, 74-83, Gurt Esparraguera and Diarte Blasco 2012, Teja Casuso and Acerbi 2010). This suburban development can be archaeologically dated to the second half of the sixth century, and was built on top of a series of loosely connected Roman suburban villas; it was finally abandoned in the ninth century, after the conquest of the city by Abd al-Rahman II during the *muwallad* revolts.

The choice of the Tagus meadow to establish the new centre of Visigothic power has been explained on various grounds, including its vicinity to the old circus, imitating palace-circus complexes from other late antique capitals (although this layout does not perfectly apply to Toledo). This opened the possibility of displaying the pow-

[9] In 1605 three public wells are recorded in Toledo, at Barrionuevo, San Salvador and Pozo Amargo (literally, 'bitter well'), and all of them are mentioned to have *aguas salobres* ('salty waters'); (Porres Martín-Cleto 1970, 113).

er of the new regime in the open fields (as opposed to the allegedly crowded upper city). The presence of the suburban churches further supports this identification. Nobody has as yet considered the possibility of direct access to water, something impossible in the upper city. Several wells and cisterns are known from the excavations, which are located very close to the river (figure 22). If there was an imitation of the imperial court ritual and display in the court of Toledo, then water surely must have played an important role in this palatine complex (baths?); something which could not be easily done in the upper city. The choice of the lower city and the availability of water can be further linked by the events which followed the Islamic conquest. The Umayyads established a fortress in the upper city, and later they built a complicated system of concatenated lifting wheels which fed water from the Tagus directly into the *alcázar*. This system was later preserved and maintained in the Castilian and Habsburg periods but it only fed the fortress, and not the city.[10] It is not possible to claim that the upper town was abandoned or completely rejected by the royal authority. In fact, King Wamba restored the gates of the city, as commemorated in an inscription (*Cont. Hisp.*, 35), the cathedral was located up there, and it is on the upper town where the Umayyads established their fortification, perhaps superimposed on some earlier Visigothic structure (Martínez Jiménez, Sastre de Diego and Tejerizo García 2018, 175-7), but the great investment in the suburb clearly make Toledo a two-foci city.

[10] These water wheels are already mentioned by Idrisi in the twelfth century (Aranda Gutiérrez, Carrobles Santos and Isabel Sánchez 1997, 138-9), and were later expanded (or repaired, or substituted) in the sixteenth century by the famous *Artificio de Juanelo*, named after Philip II's engineer Gianello Turriano. They were finally abandoned partially because the town council refused to pay for the maintenance of a water supply which only benefited the Royal palace.

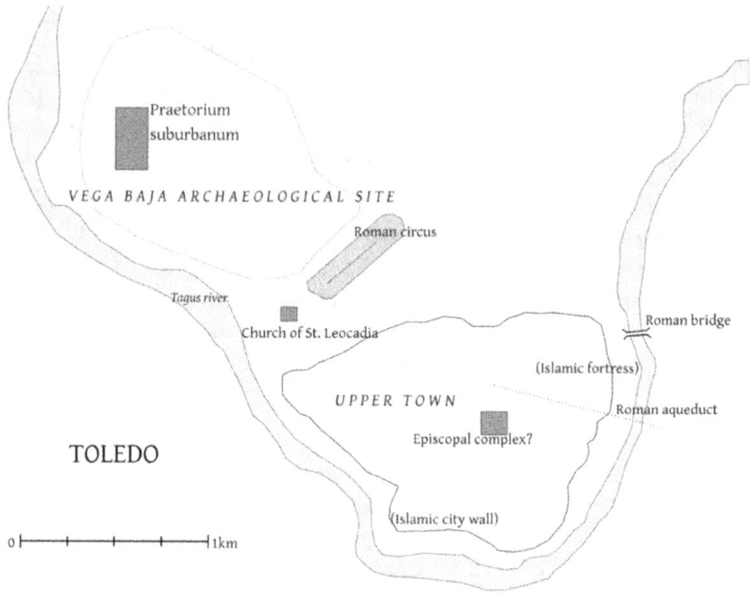

Figure 22: Reconstructed plan of Visigothic Toledo, based on recent interpretations of the remains at the Vega Baja.

Lastly, the case could be made, although without any evidence, that the reorganisation of *Italica* in Late Antiquity back to its original core (as defined by the Visigothic wall) could be a response to the end of the aqueduct. After all, the new Hadrianic expansion was developed with new branch of the aqueduct (Canto 1978), while the old colony had a pre-existing well-and-cistern supply system. The *urbs nova* is, furthermore, up on the hill, perhaps further away from the water table, repeating a situation like that of Tarragona, Toledo or Córdoba.

TRANSFORMATIONS OF THE HYDRAULIC INFRASTRUCTURE

Beyond these general considerations in the transformation of post-Roman townscapes according to the availability of aqueduct water, it is useful to have briefly a closer look at the evolution of the hydraulic infrastructures themselves inside towns. Baths, aqueducts, and conduits were reused for secondary purposes not at all related with water supply once they were abandoned, as it happened with most public buildings that lost their main function in the post-Roman period.

Baths and cisterns

As baths and cisterns went out of use, they became large empty buildings, which were reused in various different fashions. Even if most ended up being the site of dumps, some saw a secondary use as dwellings or cult buildings. However, it is important to stress that, in no case was there a conscious Christianisation of baths, or exorcisms as mentioned in eastern sources (Jiménez Sánchez and Sales Carbonell 2004, Saradi 2006, 326-7, 336, Diarte Blasco 2011, 757).

In the Iberian Peninsula there are very few examples of such Christian reuse of hydraulic infrastructure (baths or cisterns), and in most cases, there is no hint to indicate a conscious Christianisation of a bathing space as opposed to the construction of a new building (which happens to be a Christian one) on an abandoned, formerly public, structure. The first example would be the conversion of a private cistern into a *domus-ecclesiae* in Mérida during the third or fourth century, but this is both very early for our study and limited to a domestic structure. The construction of the church of Saint Genesius in Toledo on top of the aqueduct's terminal cistern mentioned above, cannot be seen as a deliberate act of Christianisation, and the cistern was surely already out of use when the church was built on top of it (Aranda Gutiérrez, Carrobles Santos and Isabel Sánchez 1997). Similarly, the conversion of one of the baths of Ampurias into a church during the Visigothic period seems to respond to this pattern of construction on an abandoned building (García-Entero 2005, 226-9).

In Seville, for instance, the terminal cistern (or one of them), excavated at the Plaza de la Pescadería was turned during the Visigothic period into a dwelling area. The presence of various hearths and dividing walls indicate this (figure 23) (García García 2007). There is some evidence to suggest a possible fifth-century phase (repairs? reconstruction?) in the *castellum* and the baths of Cartagena, but the suggestion that this had to do with a restoration of the water supply is unconvincing (Vizcaíno Sánchez 2009, 351-63). The baths of Braga were also turned into dwellings during this period (López Quiroga 2004, 75).

Figure 23: Photo of the cistern at the Plaza de la Pescadería in Seville, which was turned into a dwelling in the Visigothic period (García García 2007, fig. 9).

As far as dumps are concerned, Roman monuments would appear to have been turned into dumps during Late Antiquity. For hydraulic infrastructure this is very commonly the case, as they were large empty rooms, usually underground, which were easy of fill, and usually very conveniently located next to domestic spaces, although it is also possible that these structures were filled on purpose in order to build something on top (which could be cone with rubbish anyway). The cases of Cartagena or Toledo have just been mentioned, but the *cuniculus* of Tarragona is another good example. The cisterns and conduits of several aqueducts such as those of *Ebusus* or Tiermes are filled with plenty of fourth century material, which is very useful for dating purposes (Argente Oliver and Díaz Díaz 1980, Castro Orellana and Roig Ribas 2009, Díaz Díaz and Argente Oliver 1984). Some were used for very specific purposes, as the distribution tank of the Western aqueduct of Córdoba, which was used by a glass workshop as a dump for its refuse (Carmona Berenguer, Moreno Almenara and González Viseda 2008).

Lastly, the transformation of baths into burial areas can be mentioned. Baths were large public structures, which once they were out of use and collapsed, became relatively large areas of public land, which *technically* could be used by anyone. This may be the legal basis for the encroachment onto baths by private dwellings, as already mentioned. But this also meant that they could be turned into a burial area if no structure was built on top, as it was the case in Pamplona, where the baths, abandoned during the fifth century, were turned into an Islamic *maqbara* during the eighth (Faro Carballa, García-Barberena Unzu and Unzu Urmeneta 2008, 238-40, Unzu Urmeneta, et al. 2006, 432-3).

Secondary use of conduits

Conduits themselves were perhaps more difficult to reuse, due to their very specific elongated shape. However, because of their almost identical proportions in terms of height and width, it is very common to see conduits reused for burial purposes, as a human body could be easily fitted without much difficulty in a *specus*. In Rome, at the Crypta Balbi, the conduit was used for burials during the Gothic Wars (Manacorda 2001, 42-9), and in Spain we see late antique burials in the conduit of Tiermes and early Islamic burials inside the conduits in Mérida, at the site of Los Bodegones on the Cornalvo Aqueduct (Delgado Molina 2006).

The usurpation of the aqueducts in the countryside may be the most famous secondary use of aqueducts. Water stealing for private consumption was common in Antiquity, and as it has been already pointed out, there was a large corpus of legislation against this. Despite these laws, aqueducts have been in constant use in many places as rural irrigation channels long after they ceased to supply towns. The aqueducts of Almuñécar, Valencia, Sagunto and *Reccopolis* are known to have been in use into the twentieth century as rural irrigation channels.[11] The aqueduct of Valdepuentes however, as it passed

[11] Those of Almuñécar or Sagunto are in fact still functioning in sections as irrigation channels, and they still belong to the *comunidad de regantes* ('irri-

underneath Madinat al-Zahra, was used as the main sewer because it was five metres below the street and therefore could not supply water with pressure at ground level. Similarly, it is likely that the aqueduct of Casa Herrera in Mérida may have been used to supply the rural settlement which grew up around the basilica (Sastre de Diego and Martínez Jiménez 2012, 2013).

THE END OF AQUEDUCTS: THE END OF CLASSICAL URBANISM?

Even if the towns of Antiquity had no aqueducts at their origin, and existed for centuries before acquiring one, it is evident that after the aqueduct building 'madness' of the first and second centuries CE, towns and city-dwellers grew used to having them. We should keep in mind that by the year 400 the citizens of Córdoba, Mérida, Tarragona or Barcelona have had aqueducts and domestic piped water for almost four centuries – most of Western Europe nowadays cannot even claim two. The profound social, political and economic transformations of Late Antiquity were reflected in a constant transformation of urban layouts and townscapes. Aqueducts, as an integral aspect of urbanism, were affected by these changes.

The emergence of new urban patterns

Roman urbanism cannot be seen as something static, as it was in constant change, just as our own modern cities are. However, the image of first- and second-century Roman towns is so powerful that it is difficult to see a Roman town in any other way. The fact is that from the fourth century onwards, towns entered a different dynamic and, from then on, their evolution responded to patterns unlike those which framed the early Roman town.

Perhaps one of the most important transformations, which actually can be used to explain some of the issues proposed above, is the political reorganisation of town administration during Late Antiquity (Liebeschuetz 1992, with caveats). This readjustment substi-

gators' community') rather than to any sort of heritage bureau or department.

tuted the old traditional magistracies by the *comes* (count), appointed by the central administration, and the bishop. Town councils and local magistracies continued to exist (like the *numerarius*, or the *defensor civitatis*), but with limited powers and funds (Fernández 2017, 165-70). This meant that traditional spaces of public politics, like forums, basilicas or curias, ceased to fulfil an active function, and became mostly redundant (which can explain why they were dismantled and encroached upon). The new political situation required new forms of public architecture, and this became increasingly true in the Visigothic state formation period. This is reflected in the renovation of walls, the construction of new palaces, and the foundation of new towns, like *Reccopolis*, El Tolmo, the Suevic acropolis of Falperra outside Braga, or the obscure and unknown Visigothic foundations of *Victoriacum* and *Ologicus*. These constructions not only benefited from the direct involvement of the monarchy, but also prompted the development of a new service elite which further developed a post-Roman architecture of power. This process of transformation continued during the Umayyad period, when a yet further political reorganisation prompted the creation of an utterly new architecture of power (Martínez Jiménez, Sastre de Diego and Tejerizo García 2018, 276-85).

Besides these political transformations, which affected the evolution of public architecture, a key factor to consider are religious transformations: from pagan to Christian and then to Islamic. The transformation of religious space did not only affect cult buildings, but also their environments. This is especially true for the evolution of Christian cult centres, which in the early stages of urban Christianisation were located outside the city walls, in the burial areas linked to a martyr's grave. These could, in the end, develop into a suburb, as in Tarragona, Córdoba, Metz and Tours. Linked to the development of new architectures of power, the construction of episcopal complexes is one of the key developments in public monumentality, because these became the new, visible centres of political and religious power (Miller 2000). They appear in Barcelona (Bonnet and Beltrán de Heredia 2002), Tarragona (López Vilar 2006), and *Egara*,[12] and can be inferred in Valencia, Córdoba and Mérida (Sastre de Diego 2015). Furthermore, Christian constructions and other pious acts were the only possible niche in which old euergetic practices could be displayed,[13] something which has been much more thoroughly studied for Italy and Merovingian Francia (Beaujard 1996, Guyon 2006, B. Ward-Perkins 1984, Loseby 2006).

Lastly, and imposed by the historical context, the economic changes of Late Antiquity and the early Middle Ages should not be underestimated. As the Roman political world disintegrated, so did its economic sphere of influence. The interconnections between regions dramatically decreased in volume, and the highly specialised systems of production which relied on long distance exchange collapsed to regional or local patterns (B. Ward-Perkins 2005, Wickham 2005). This meant that gradually the different regions became increasingly isolated from each other. Of course, trade continued, and exceptional sites such as Toledo, *Reccopolis* or Mérida show signs of continuing long distance trade and production of goods. Coastal towns from *Sala* in Morocco to Vigo and (of course) the Mediterranean coast kept close trading links between them and, during the period of Byzantine occupation, well connected with the East, although this never translated into a wide distribution to their hinterlands (Bernal Casasola 2008, Fernández Fernández 2014). But this slowly faded through the sixth and seventh centuries. The final disintegration of the late Roman models, as continued by the Visigothic kingdom, occurred during the eighth century and under the Umayyad regime; the Umayyads reactivated trade and production, but on completely different grounds. These may have included Mediterranean links, but with different goods, routes and main trading nodes.

In this context, slow transformations of the late Roman world, which lasted for several centuries, the Roman townscape changed, attracted by the new foci: the new centres of power and religious

---

[12] Even if there was no urban centre in *Egara* (Oller Guzmán 2014).

[13] One particularly late euergete is the *vir inlustris* Gudiliuva, who dedicated three churches in the early seventh century: *ICERV* 303.

complexes which developed mostly during the sixth century. However, from the fifth century onwards there had been a trend to reduced wall enclosures, resulting in smaller towns, parallel to the development of suburbs around the functioning harbours or emerging religious centres. Paradoxically, towns were becoming smaller, but yet spreading over larger areas. These are processes which took place all over the West (Esmonde Cleary 2013). Perhaps spreading the population in this fashion was another way of maximising the water resources.

Public spectacle monuments such as theatres, circuses, and amphitheatres (despite some dubious sixth-century mentions) ceased to function in this period. Many had already been quarried away in the fourth century, as in Córdoba, or else turned to areas of Christian worship, as in Tarragona. These buildings were deeply rooted in earlier Roman cultural and religious traditions and interlinked with old ideas on public munificence and obtaining political benefit from investing in public displays, something which had faded away by the end of the Roman Empire.

The abandonment of forums, basilicas, and theatres implies a deep cultural change in town dwellers, probably derived from the social, economic, political and religious changes which characterised Late Antiquity. A new concept of town emerged as a result of these transformations, although it still was based on a symbolic, administrative and economic importance. In the classical Roman world, baths and aqueducts were an inherent part of culture, so what can archaeology tell us about the relevance of both in this period of transformation and beyond?[14]

As has been stated, aqueducts were not originally an urban need (with only a few exceptions) and they were linked to the development in the provinces of Roman bathing culture. Of course, the de-

[14] There are many references in the classical authors to baths and the 'Roman way of life', and almost as many on aqueducts; perhaps the most significant is *CIL* VI 15258: *Balnea, vina, Venus corrumpunt corpora nostra; sed vitam faciunt balnea, vina, Venus* = 'Baths, wines and love corrupt our bodies, but baths, wines and love give [us] life'.

velopment of long-distance aqueduct supply enabled the settlement of areas which would have been discarded as not ideal in earlier periods. In fact, even if they were originally not a basic foundation stone of towns, aqueducts became over the centuries an integral part of urban settlements, permitting several areas to further develop and to grow, which is evident only after these systems ceased to function. As far as bathing is concerned, the changing attitude towards classical bathing was modified by the new Christian moralities and the decay of the infrastructure marked the end of Roman bath culture.

Were aqueducts necessary in the end?

It is very difficult to assess whether aqueducts were a truly urban need or simply an extravagant luxury inherited from the Roman past from our evidence. It would be anyway foolish to answer this question with a simple yes/no, black/white answer, and perhaps it would be wiser to answer on a case-to-case basis. However, for the sake of a possible model or pattern, some overall conclusions can be drawn from the archaeological evidence.

Aqueducts, as we have seen, enabled the development of areas which previously could not have been densely urbanised, and for a long time these conduits were responsible for the cohesive nature of many urban nuclei. Aqueducts therefore acted as a key element in town planning and urban development, especially in the early imperial period. Furthermore, they prompted the construction of public bath complexes, and public fountains, and greatly contributed to the development of Roman *domus* in the provinces. The very close relationships between these developments, which are all roughly contemporary in their origins, and long-distance water supply systems, are really noticeable, paradoxically, once the latter ceased to function. After the end of aqueduct supplies we see a return to pre-Roman urban patterns, not only in terms of water supply (wells and cisterns), but also in terms of settlement location. The reorganisation of urban settlements according to the availability of water is one of the aspects which characterises later medieval urbanism into smaller more concentrated areas (even leading to poli-nucleated settlements). It can also be argued that the decline of public water supply systems went hand in hand with a decline in the number of inhabitants which could live in a given town.

In this way, aqueducts were an active component of the 'Roman way of life' in the provinces, especially with regards to public amenities. Ongoing aqueducts in Late Antiquity were then monuments which promoted or sustained traditional Roman urban life, especially at a privileged level. After all, the old municipal elites had been responsible for the construction of this infrastructure, and were the ones that benefited most from it (Fuentes Domínguez 2000, 138). Aqueducts were, therefore, not only a vehicle for urban concentration in the Roman period, but also a means of expressing urban elite behaviour. In Late Antiquity, it was necessary for these elites to adapt to the new circumstances and to come up with alternative ways of displaying their status.

There is a symbolic meaning attached to aqueducts, beyond the purely materialistic and tangible benefits for urban development. This perhaps can further explain the continuity of some aqueducts, and the development of new urban patterns, and will be explored in the next chapter.

AQUEDUCTS AND THEIR SOCIO-POLITICAL IMPORTANCE IN LATE ANTIQUITY

Beyond the architectural and archaeological impact aqueducts had, it is important to remember that they were built by and for town dwellers; throughout the Roman period the people co-inhabited with aqueducts and benefitted from their water. This was, in many cases, a very long-lasting relationship from a functional point of view, which greatly modified urban life. Nevertheless, aqueducts also had a very important symbolic meaning, as they were not only a source of pride for those who built them, but also for those benefitting from them many years later. Aqueducts were part of Roman urban pride, and this feeling went beyond the urban elites to the higher echelons of the state and the Church.

It has already been discussed how the continuity of aqueducts affected the urban landscape, and now we will focus on another main aspect: the relation between town dwellers, urban elites and the state (both the Visigothic and the Umayyad) with aqueducts.

AQUEDUCTS AND THE URBAN ELITES

For fourth-century urban dwellers, aqueducts may have seemed eternal. As opposed to walls (which were in many cases being built or remodelled in this period) or old forums (which were slowly but constantly going out of use and dismantled), aqueducts stood, had been standing seemingly for ever and were still functioning. As has become obvious from previous chapters, soon this situation would change. The previous chapter has explored how this affected the development and evolution of urban areas in terms of settlement location and spaces of power, but it is important to keep in mind that urban dwellers (and especially, the urban elite) were directly involved in the evolution of aqueduct supply systems.

An elite water supply system?

Aqueducts were elite buildings, built by the elite as part of early Roman euergetic practices, and were closely linked to curial administration, so the fate of urban elites and urban aqueducts would appear to go hand in hand (Martínez Jiménez 2019). In fact, their evolution is so closely related that the origins and decline of one of them could be used as a perfect analogy to explain the transformations of the other. This comparison is most clearly exemplified by the chronological correlation between the decline of the power of curial municipal administration and the beginning of the end of aqueducts (from the fourth to the sixth and seventh centuries), and firmly based on the nature of aqueduct maintenance and repairs.[1]

This close correlation may also be the reason behind the continuity of certain aqueducts, and it is clear from the evidence put forward that aqueducts continued mostly in those towns which preserved some degree of administrative and political importance and of commitment towards the community. It would not be fair to claim that Roman water culture was exclusively an elite culture, but it was elite-led, as the members of the elite, following Vitruvian canons, received in terms of private domestic supply a third of the town's aqueduct water.[2] Most of this water was used for private baths and fountains.

---

[1] Referring to Roman culture in general, J. Matthews (1990 [1975], 328) mentions how '[y]et it was in the hands of such men [i.e., the elites], in their enterprise and willingness to play their part in the new conditions of the fifth century that the survival of Roman culture in the west resided', which can perfectly be applied to aqueducts.

[2] *Arch.* VIII.6.2: *Ita in medio ponentur fistulae in omnes lacus et salientes, ex altero in balneas vectigal quotannis populo praestent, ex quibus tertio in domus privatas, ne desit in publico* = '[In the castellum] from the middle receptacle pipes will be taken to all pools and fountains; from the second receptacle to the baths, in order to furnish a public revenue; to avoid a deficiency in the public supply, private houses are to be supplied from the

Urban aristocrats enjoyed bathing as much as the urban plebs, and many members of the elite had their own private baths. This was only possible after the development of aqueducts, and became an essential part of Roman water culture. In fact, the fourth century saw an increase in the number of private baths. However, in Late Antiquity the failing aqueducts threatened to put this practice to an end. In some instances, such as in Barcelona, the continuity of private baths into the fifth and sixth centuries (those of Pati Llimona or Bisbe Caçador) was possible due to the continuity of the aqueduct. The private baths of Seville dated up to the fifth/sixth centuries at Cuesta del Rosario and the Palacio Arzobispal excavations are further examples of this (González Acuña 2011, 487, 492). Seville and Barcelona are two good examples of towns with administrative or commercial importance and a strong elite, in which aqueducts and private baths continue into this period.

However, there are instances of similarly important towns where aqueduct supply was not reliable (or completely gone) in the sixth century, as in Tarragona and Mérida. There, the members of the elite looked for alternatives that would allow them to preserve their aristocratic domestic baths, despite the end of aqueduct supply. The end of aqueducts meant that wells became the main source of water: the various late baths of the harbour and Christian suburbs in Tarragona (the fifth-century baths at the Parque Central and the fourth-to-seventh century houses of PERI-2) were all supplied by underground water. The late suburban private baths of Mérida (Calle Nerja excavations, and Resti suburb) were also fed by water from wells, sometimes linked to water-wheels. Note that baths fed by wells were not late antique innovations, as pre-aqueduct baths at Pompeii were fed by similar mechanisms. Neither were they particular to Spain. However, this extra effort and search for water was not always possible, and Mérida offers further examples of this process of

---

third'. Vitruvius' description of a *physical* tripartite division is misleading, but the truth behind this is that private consumers, after paying a fee, were entitled to a private connection to the main water distribution network (Brunn 2002b).

adaptation. For instance, at the site of Morería the subdivision of the *domus* into various tenant houses meant that the old *balneum* was also turned into a dwelling area. Similarly, in fifth-century Gaul, Sidonius Apollinaris mentions that 'both of his guests had baths being built, but none in use',[3] showing how private bath-building initiative (even if not always successful) continued.

The large corpus of evidence for private late antique baths available for the Iberian Peninsula compiled by Virginia García-Entero in her PhD thesis (García-Entero 2005) confirms that this latter process (abandonment rather than adaptation) is far more common than the continuity outlined above.

Civic duties

Aqueducts were perhaps elements of the elite Roman culture, but they were also fulfilling a public service, although in Late Antiquity it is difficult to assess the extent to which elites were willing to preserve them.

The transformation of the urban administration in *Hispania* (Curchin 2014, Fernández 2017) despite the late antique links between the large landowning aristocracy (the urban elites) and the central government meant that traditional euergetic practices were increasingly abandoned from the fourth century onwards. Especially after the fifth century. The members of the civic councils, the *curiales*, had lost most political incentives to invest in public munificence and, in many cases, lacked the resources to carry out these constructions. Baths, fountains, circuses, and aqueducts, were repaired during the late Roman period by governors and other members of the higher echelons of the administration, linked to the imperial networks of patronage. These were the persons who had both funds and civic duties to fulfil (as we see in the baths and fountains of Tarragona and Lisbon or the theatre and amphitheatres of Mérida). During the sixth century, Visigothic laws highlighted the importance of preserving *civilia munera*, continuing late Roman legislation. This was most

[3] Sid. Ap., *Epist.*, II.2.8: *Balneas habebat in opere uterque hospes, in usu neuter.*

certainly linked to the role of the *curiales* as tax collectors, although the maintenance of buildings and supervision of repairs might also have been implied.

The preservation of Roman town life by urban elites in a few chosen sites is evident, and the preservation of aqueducts, or the attempts to keep them running, is one sign of this. There is some evidence to suggest that these attempts were not simply done for the sake of the elite's own private *balnea*, but with a public purpose in mind, which would indicate a late continuity of munificence or, at least, concern for the community. In this context, the implied mention to curial offices, secondary to the count (the royal figure of authority and representative of the central administration), in sixth-century addenda to some Roman laws (Curchin 2014) could be related to this post-Roman administrative structure. The construction of a new set of public baths behind the theatre in Mérida (García-Entero 2005, 527-9)[4] is very interesting in this respect, which could be paralleled to the obscure mention to 'circus' games in Zaragoza in the sixth century (*Chr. Caesaraug.*, 504), but on a much more solid archaeological basis. These baths (figure 24) were fed by a well, but their size and location do not suggest a private structure. This could perhaps indicate an attempt to continue with traditional euergetic practices, despite the lack of an aqueduct. Anyways, this seems to be a rather isolated example: traditional Roman munificence practices had declined amongst urban aristocrats, who now chose to show their power and wealth by building churches or private houses, a clear reflection of the rupture between Classical and late antique practices (Kulikowski 2004, 70). Besides those of Mérida and those of Barcelona (further discussed below), there are no other known public baths functioning in this period, not to mention newly-built ones. Even so, by the seventh century, church donations by a layman, such as those of *vir inlustris* Gudiliuva, were already equally rare.[5]

---

[4] Although not all the results have been published. See further discussion below.

[5] *ICERV* 303 commemorates the construction of three churches, allegedly in rural contexts: *[In nomin]e d(e)i n(o)s(tr)i Ih(e)s(u) Chr(ist)i consecrata est*

Figure 24: Sixth-century new set of baths built behind the theatre peristyle, in Mérida. Note how high the baths are built in comparison to the original Roman street level. These baths were supplied by wells and had small individual tubs rather than communal pools.

As far as aqueducts and water supply are concerned, if imperial law (in particular *CTheod* XV.2.1 = *CIust* XI.43.1.1) is a reflection of provincial practices, then we have to assume that it was in the hands of private landowners to repair aqueducts once the municipal administrations had given up their maintenance. Again, it is difficult to say up to what point landowners were willing to take over this responsibility in exchange for tax deductions, especially after the fifth, century when the collection of taxes was certainly in radical decline, and by

[ec]clesia S(an)c(t)i Stefani primi martyris in locum Nativola ... ab inl(ustre) Gudiliu[va] cum operarios vernolos et sumptu proprio = 'In the name of God our Lord Jesus Christ, this church was consecrated at the site of Nativola to Saint Stephen, first martyr ... by [vir] inlustris Gudiliuva with his own slave-workers and money'. Of course this may just be a decline in the epigraphic habit, and the lack of charters for this period prevents any further guesses on private church donation (B. Ward-Perkins 1984, ch. 2).

the Visigothic period of state formation, when taxation was partially reintroduced (Martín Viso 2013). It was really up to their euergetic disposition (or their private interest) to preserve aqueducts running.

Social and cultural redefinitions?

From our modern perspective, it seems inconceivable that the municipal authorities or the urban elites would be ready to let aqueducts fall to ruin, but considering the evidence presented, this could have been the case. Despite some very late and exceptional cases of public euergetism, by the fifth and sixth centuries urban elites lacked any incentive to continue with these anachronistic practices, and they rather invest in the construction of churches. The lack of technical capabilities greatly limited their ability to repair aqueducts, but also the decreasing power and impoverishment of municipal infrastructures meant that town councils had to rely on other patrons to preserve them. Normally these would have been provincial governors (cf. *DLH* II.20) or the emperor himself, but after the fifth century this was no longer an option. Despite any urban pride and interest the elites may have had, some things were simply beyond their control (Greenhalgh 1989, 110).

Important towns enjoyed a longer period of aqueduct supply, although in most cases (Barcelona, Valencia, Tarragona, *Reccopolis*, and perhaps Córdoba) either the state or the Church took over the aqueducts.

But overall, it can be said that direct access to water was becoming increasingly a sign of higher and higher status, even more than in previous periods, due to its scarcity. The declining number of functioning aqueducts also prompted changes in more general (non-elite) water culture, even if it was simply because large public baths were no longer available, excepting clerical baths, which may have been of public access, as known for Italy (B. Ward-Perkins 1984, 135-41). Bathing was reduced to the domestic space, if preserved at all, or focused on the new Christian centres. The expansion of new Christian morals, which had a lot to say on Roman bathing,[6] may be one of the reasons behind this, but we should not take the texts of Christian moralists at face value, as bathing habits may have still been desirable (Miller 2000, 32).

It was not simply that the infrastructure was declining, that the new Christian ambience seemed opposed to traditional bathing, or that the elites could not take care of the old baths (Osland 2011, 361); one of the key issues is that one of the results of the transformations of Late Antiquity is the transformation of the elites. From the fifth century onwards, they lacked any incentive to preserve traditional public munificence, as they were further and further away from the public municipal administration and the Roman state that had created it. During the period of a power vacuum of the fifth century, but especially through the process of state formation of the Visigothic kingdom, the elites had to accommodate to the new situation. As has been said already, there was an almost total rupture between urban/local affairs and the state (excepting in those areas of direct control, such as El Tolmo, *Reccopolis*, or Toledo), and the only way the local elites had to achieve rank and position was to turn into a service aristocracy. Serving in the public administration (regardless of ethnicity or religion) was the way to obtain power, but this new administration was far removed from municipal issues: it had been the new way forward in the fourth century, and by the sixth century it became a must (Fernández Delgado, Martínez Jiménez and Tejerizo García 2013).

This in turn gives another reason for the decreasing interest in preserving the aqueducts: The Visigothic state (the elites' model to follow?) itself did not take much care in aqueducts.

[6] Especially regarding mixed bathing, the pagan connotations of the statues in baths, the perception of bathing as a decadent luxury, and the relationship between bathhouses and prostitution.

THE RELEVANCE OF AQUEDUCTS FOR THE VISIGOTHIC MONARCHY

As far as the Visigothic monarchy was concerned, public monuments were largely a responsibility of the local administration. As opposed to what the Ostrogoths (Theoderic in particular) did in Rome, the Visigothic kings hardly cared for public services or monuments. This is perhaps not simply a matter of lack of interest, but rather a consequence of the differences between early sixth-century Italy and late sixth-century *Hispania*. Yet, in this context, we see the construction of a new aqueduct at *Reccopolis*.

Germanic monarchies and public works

Cassiodorus in his *Variae* mentions how the old public monuments were an element of pride for the state, and how the pride of the Goths was to preserve the Roman way of life (*Var.*, IX.14.18: *laus Gothorum est civilitas custodita*). The concept of preserving the past, the Roman order, and the repair of old buildings served Theoderic as a way of presenting himself as a legitimate ruler of Italy, which in turn was sanctioned by the Ostrogothic legislation on this topic, also compiled in the *Variae*.[7] The corpus of repairs on monuments, especially in (but not limited to) Rome and Ravenna, and new constructions is vast. This attempt to promote the image of the Ostrogothic monarchs as continuators of the Roman order was closely linked to the preservation of monuments, because in Italy the emperor had been much closer and much more involved in this type of works than in the provinces. The clearest way to show continuity and the lack of interruption between the imperial administration and the

[7] Many passages of Cassiodorus refer to Theoderic's policy as a righteous cause: *Var.*, I.25.1: *Nil prodest initia rei solidare, si valebit praesumptio ordinata destruere, … de custoditis adquiritur laudata perfectio* = 'It is not useful to build firmly from the beginning if lawlessness has the power to ruin what has been designed, … from the things preserved the glory of completion is acquired'; III.30.1: *quid est enim dignius, quod tractare debeamus, quam eius reparationem exigere, quae ornatum constat nostrae rei publicae continere?* = 'What indeed is worthier, than to maintain the repairs of that place [Rome] which clearly preserves the glory for our government?'.

Ostrogothic kingdom was maintenance and constructions. Furthermore, in Italy all the imperial administrative infrastructure was still very present, so the transition did not imply great changes, as the formulae preserved in the *Variae* would indicate.

In this aspect, *Hispania* was very dissimilar to Italy, and not only because there is no corpus of letters comparable to the *Variae*. By the time the Visigoths had firmly established and reorganised the kingdom (especially under Liuvigild and Reccared, in the last quarter of the sixth century) most of the Roman infrastructure had long disappeared. In fact, even if the Visigothic administration imitates late Roman practices, it had to be mostly reinvented and restored. Looking back at the Roman past was a very useful legitimising argument used by the Visigoths, but the scope of their public constructions was much reduced, so the process of state formation was based overall in less tangible things, like court ceremonial, law giving, and tax reorganisation.[8]

When it comes to aqueducts, the Ostrogoths again were much more involved in their maintenance. As we have seen for Rome and Ravenna. The Visigoths, on the other hand, were only responsible for the construction of the aqueduct of *Reccopolis*. Perhaps the failed restorations of the aqueduct of Mérida (figure 9) could be related to a royal project; after all, Mérida was for many years not only the capital of the kingdom (Ripoll López 2000), but also the most important city, and the Ostrogoths, who were keen on repairing aqueducts, must have been a model to follow. Furthermore, we know of other royal constructions through inscriptions found in Mérida, even if of dubious nature.[9] However, this is only a supposition, as it is equally

[8] Cf. *Hist. Goth.*, 51. In fact, all of this may be an imitation of the contemporaneous *Eastern* Empire rather than a revitalization of the old, long-gone, Western Empire. This is further discussed by M. R. Valverde Castro (2000, passim, esp. pp. 141-76, 182-94).

[9] The mysterious inscription by King Chindaswinth (*ICERV* 366: *[Chi]ndasvintus rex pi[issi/mu]s in <i>mperio/ [Ch]indasvintus [...]*), preserved with no context, is generally considered to have been linked to a building.

possible that the local elites or the bishops were responsible for this attempt.

Restoration of public works

The repairs carried out by the Visigothic monarchy are very few. In fact, they are so few that the examples should be considered exceptions for various reasons. Perhaps the reconstruction of the bridge and walls of Mérida in the late fifth century (*ICERV* 363) could be considered the first example, and even here, the intervention of King Euric is a pure formality, as the initiative derived from the bishop and the *dux* (Arce Martínez 2008). After these, the restorations of the walls of *Italica* by Liuvigild (Bicl., 584.1), together with the repairs of the walls of Toledo later in the seventh century by Wamba, are perhaps the best known royally-sponsored reconstructions (*Cont. Hisp.*, 35).[10] However, these are not restored public services but walls, which are very different in their nature, and it should be taken into account that the former were carried out during the rebellion of Hermengild, who was stationed in Seville, and the latter belong to the capital city. The inscription of Chindaswinth in Mérida mentioned above and the inscription of Hermengild in Seville could also be related to (re)construction of buildings, but nothing is known about them (Fernández Martínez and Gómez Pallarés 2001).[11] Restoration of walls was a very important and symbolic thing to do, not

---

[10] Known because of a commemorative inscription, now lost, about the repairs of the walls of Toledo: *Erexit, fautore Deo, rex inclytus urbem/ Wamba, suae celebrem protendens gentis honorem.//Vos, sancti Domini, quorum hic praesentia fulget/ hanc urbem et plebem solito servate favore* = 'With God as patron, the famous King Wamba erected the city to increase the honour (and) fame of his people. You, holy Masters, whose presence shines here serve this city and (its) people with your customary favour'.

[11] *ICERV* 364: (Chrism) *In nomine Domini ann[o f]eliciter secundo regni Dom(i)/ni nostri <H>erminigildi regis quem persequitur genetor/ su(u)s Dom(inus) Liuvigildus rex(.) in cibitate(m) <H>ispa(lim) ducti aione* = 'In the name of the Lord, in the second year of the happy reign of our lord King Hermengild, whom his father our lord King Liuvigild persecutes. Brought (=Hermengild) into the city of Seville for ever (*aione*)'.

only because city walls had become one of the most important urban symbols (in words of Isidore: *urbs ipsa moenia sunt* = 'a city itself *is* the walls', *Etym.*, XV.2.1), but also because they are very visible, and practical. It was a way to make strong statements about protecting a city.

This is how far the Visigothic monarchy went with regards to public services. There were other public buildings and civic constructions built in the Roman fashion by the monarchy (new palaces in Córdoba and Toledo; new urban foundations such as El Tolmo, *Reccopolis*, and *Victoriacum*) linked to the public administration, but these can hardly be considered to have provided public services; even if they were still necessary buildings for the new government (Ripoll López 2000, 374, Valverde Castro 2000, 182-3). This contrasts greatly with Ostrogothic Italy, but is perhaps close to what is known in Merovingian Francia. In Vandal Africa baths seem to have been repaired too, alongside other public buildings, but just as it is the case with Ostrogothic Italy, the transition between the late Roman and the post-Roman administration was very short, and hardly suffered any disruption. Gaul and Spain, however, suffered periods of power vacuum which badly damaged the administration and its reach.[12]

Besides this, and as it has been already highlighted earlier, the lack of skilled engineers and the loss of vaulting techniques probably meant that by the late sixth century it was technologically impossible to repair most public buildings, regardless of the interest shown by the administration (Utrero Agudo 2006, Martínez Jiménez and González Gutiérrez 2017). The reconstruction of walls requires only a basic knowledge of engineering and construction; repairing a set of vaulted baths or creating a levelled conduit for an aqueduct was not that easy. As for the construction of the large new royal projects (*Reccopolis*, the *praetorium suburbanum* of Toledo, etc.), specialised builders and engineers could have been mustered from abroad, as suggested earlier.

---

[12] For the Vandal examples, see the poem *De Thermis Alianarum* by Felix (*vir clarissimus*), which praises a reconstruction in particular by King Thrasamund (Chalon, et al. 1985).

Overall, it would be misleading to take Ostrogothic Italy or Vandal Africa as a comparison with Visigothic Spain when it comes to the maintenance of public buildings and the importance given to aqueducts. It is perfectly understandable why the Visigoths might have wanted to promote hydraulic architecture and to preserve aqueduct supply systems, but various circumstances prevented this from happening.

The aqueduct of *Reccopolis*

In this context, the aqueduct of *Reccopolis* stands out as a unique exception. It is my opinion that it was the work of foreign engineers, brought with the purpose of building *Reccopolis*. Its importance as an element of prestige has been discussed in print already, but the fact that this is an aqueduct built in the late sixth century in the West deserves more than the usual four lines (Ripoll López 2000, 393, Ripoll López and Velázquez Soriano 2008, 217).

The aqueduct was, from a landscape analysis, a very visible monumental building, and it was also a statement of the control the town exercised over its territory and the confidence of the builder that the surroundings were safe. It was a display of power over nature and a way of showing off the resources available to the monarchy, especially in a context where such monuments could not normally be built any more. But perhaps, above all, to build an aqueduct was a way to refer back to Roman urban foundations imitating Roman imperial practices of public munificence; an aqueduct in *Reccopolis*, as part of the ideal town, was a must. The foundation of *Justiniana Prima* by Justinian in the Balkans is a good comparison, because just as *Reccopolis*, it has a new aqueduct (Ivanišević 2012, 2017). The aqueduct of *Justiniana Prima* brought water from over 20km to the lower town (compare it to the 5km of the *Reccopolis* aqueduct), where the baths are located, whereas the upper town and the acropolis seem to have been supplied by cisterns. A new urban imperial foundation, despite its small size, required an aqueduct to be fully considered a city.

Furthermore, aqueducts (as well as walls, which *Reccopolis* also had), still had a functional public purpose in late antique towns. It is perhaps because of this that the Ostrogoths and the Vandals took care of the maintenance of aqueducts: they were clearly visible, easily recognisable, and evidently useful – and not simply because it was another imperial practice they could imitate. Linking the ruler's name to one of these structures was a great act of propaganda while still providing a public service (and one which was still in demand), so it was a perfect investment. Aqueducts were, after all, standing wonders of engineering.[13]

The aqueduct of *Reccopolis* has to be considered in the context of the new urban foundation, a personal project of King Liuvigild, which drew large amounts of effort and resources. Even if preserving and restoring aqueducts in other towns (maybe in Toledo, the capital) might have given Liuvigild popular recognition, the truth is that by this period Visigothic kings did not need popular support to legitimise their rule. By the late sixth century, Visigothic power was based on the emerging service elite, the mix of ecclesiastical, aristocratic landowning, and military elites, who earned their status by rank and not necessarily by birth or ethnicity (Buchberger 2013, Fernández Delgado, Martínez Jiménez and Tejerizo García 2013).

Even if it might have been possible for the monarchy to spend resources in restoring other aqueducts in the Peninsula, to further legitimise its position, there was no demand for these repairs to be carried out. The resources and skilled workmen might have been temporarily available, but, all their efforts were focused on the two main royal projects (*Reccopolis* and the Vega Baja suburb in Toledo). This is parallel to the works of the Ostrogoths, who focused most of their efforts on aqueducts linked to their own royal residences, rather than on new projects. By the time of the Visigothic state formation, the dissociation between central administration and local monuments had reached a point of no return, and only when royal power had to be displayed was a new administrative building set up.

Water euergetism?

Considering the repairs and constructions commissioned by the Visigothic monarchy (which are few) and the exceptional example of the aqueduct of *Reccopolis* (which should not be considered as water eu-

[13] Cf. Frontinus (*Aq.*, I.16), quoted above.

ergetism on its own, but rather part of an overall scheme), Visigothic water euergetism is noteworthy for its absence. As I have tried to put forward, it was impossible for the Visigothic central administration to carry on with traditional euergetic practices beyond walls and churches.[14] In fact, the only other water-related dedications by the monarchy are also church-related: the medicinal water springs close to the church of Saint John in Baños mentioned earlier and the monastic complex built next to the new royally-sponsored monastic complex of Guarrazar (Pérex Agorreta and Miró Alaix 2018, 285, 365). Similarly, later on, in the ninth century the kings of Asturias would build an aedicule to hold a spring, a sort of public water supply in their newly-built capital of Oviedo (figure 11), but this was far removed from Roman traditional practices, and perhaps it is not until the fifteenth century when the kings of Castile make laws to preserve the aqueduct of Segovia that we will see the state intervening again in a major way with urban water supply systems.

THE CHURCH AND URBAN WATER SUPPLY SYSTEMS

The Church was one of the most important institutions in Late Antiquity, and its influence in towns and urban matters increased throughout the period: the importance of the Church in towns, channelled through the actions of bishops, grew from the fourth century onwards, so that by the sixth century, these men were largely in control of municipal issues. Similarly, Christianity and the Church had a very deep impact in the transformation of traditional Roman social values and attitudes to old Roman ways. Baths, as major consumers of aqueduct water, are key to understanding the transformations of urban water supplies. These two things put together are important for this study, because the new Christian setting did not favour traditional Roman bathing habits or bath buildings, and

[14] And, furthermore, the Visigothic monarchy had other immediate priorities linked to its process of state formation and the object of its investment of resources and capital (cities, walls, diplomatic exchanges, gift giving). The munificence strategy was different to that of the Roman period.

bishops had the means and power to reformulate the priorities of water supply.

The new urban leaders

The evolution of the bishops' power in the late antique town was a long process, which started with Constantine's concessions in the early fourth century (Matthews 1990 [1975], 307). Their growth in relevance and importance was such that by the sixth century, they had become the visible leaders of local urban communities, and biased ecclesiastical sources such as the works of Gregory of Tours and the *Lives of the Fathers of Mérida* present this. Similarly, the virtual disintegration of urban magistracies (not of the *curiales* themselves) left aqueducts and other public buildings apparently unprotected. Imperial legislation had these monuments preserved as public property, but at a town level, bishops, who were the only acting local power, might have put themselves forward as keepers of these monuments.

Archaeology reflects this transition, and the episcopal complexes of Barcelona, Valencia, Tarragona and Mérida have already been mentioned as examples of the new urban order. Episcopal complexes, usually referred to as *episcopia*, were the monumental reflection of the bishops' power in their cities. Originally an evolution of aristocratic houses, the *episcopium* was in itself by the sixth century a set of buildings, which included the cathedral basilica, a baptistery and residential quarters, which were also used for administrative and business purposes (Miller 2000, 13-20).

When it comes to aqueducts, it seems clear from Italian examples that bishops became highly involved in the maintenance and preservation of aqueducts: in Nola (Paul. Nol., *Carm.*, XXI.704-53), the local elites allowed the bishop early in the fifth century to build an aqueduct, which they later regretted, as the bishops ultimately monopolised the use and maintenance of this supply.[15] In Naples (B. Ward-Perkins 1984, 131-2), the bishop had also monopolised the aq-

[15] '[B]ishops subordinated the requirements of their communities to those of their churches'; Squatriti 1998: 12-7 (quoted, p. 13).

ueduct, taking away the responsibility from the municipal authorities, as shown in a letter by Pope Gregory (*Epist.*, IX.76).[16] Water still used to show high standards of living, a sign of luxury, and a display of the power of the Church, much in the way old urban aristocrats had done with their private baths. Episcopal patronage over the water supply was usually linked to a new Christian charity, bringing water for the Church baths, fountains, and to ecclesiastical complexes.

In the Iberian Peninsula there is no direct evidence to suggest this, apart from the episcopal baths of Barcelona. Similarly, the conduit at Casa Herrera may have been for the general use of the (ecclesiastical) settlement (Sastre de Diego and Martínez Jiménez 2013). Besides this, what we find is that aqueduct supply was preserved in the immediate environments of episcopal complexes, as in Tarragona, Valencia, Barcelona, Seville (Martínez Jiménez 2012),[17] although this cannot be positively linked to ecclesiastical maintenance due to the lack of written or epigraphic sources. Nonetheless, it would not be wise to assume that even after these consciously-located constructions bishops did not become much more actively involved with water supply systems (with some degree of agreement with the municipal administration). Christian charity and old traditional munificence did slightly overlap, and there may be some political intention behind the construction of episcopal complexes in the immediate environments of sources of water. Especially if, as shown, public water supplies were becoming less and less frequent. This would have helped bishops to reinforce their position within the community by controlling the distribution of water, and ideally, controlling its basic maintenance as well.

---

[16] We know that it was out of use during the Gothic Wars (*Bell. Goth.*, I.9.10-8), so either it was repaired or the letter refers to a new conduit.

[17] I originally proposed that Córdoba was another example. But this has been since disproven (Ruiz Bueno 2018b, 58-60).

Baths and bishops

The decline of Roman traditional bathing throughout this period may be one of the reasons behind the declining interest in preserving aqueducts. Baths were the main consumers of aqueduct water, and many conduits were built simply to supply sets of baths. With the spread of Christian morality traditional baths faded away and aqueducts lost one of their main consumers.

The position of the Church towards bathing was ambivalent, to say the least. While therapeutic bathing was certainly accepted by the Church Fathers, Roman social and luxurious bathing was against the ecclesiastical position regarding Christian spirituality and the negation of luxury. It should be noted too that despite this, the Church never forbade or publicly rejected bathing (Yegül 1991, 314-8). Even if the Church cannot be responsible for the closure of public baths, it is true that Christian moralists promoted a new type of bathing which eventually led to the disappearance of Roman public baths. For these moralists, bathing was to be encouraged mostly as a healing remedy, and always in individual tubs, something which could be done in small baths without large pools, and which could even be filled from wells or cisterns with buckets. Although no examples are known for Spain, this was commonplace in eighth and ninth century monasteries in Italy (Magnusson and Squatriti 2000, 246). In Gaul this was the case as well, although in Saint Denis a new water conduit was built from the 'holy spring' to bring water to the bathing area (Benoit and Rouillard 2000, 167-8, Wyss 1996, 188-9). It is difficult, nevertheless, to assert whether the new Christian bathing culture was a result of the decreasing number of available public baths, or if the newly imposed morality was a trigger for the further abandonment of buildings which were already suffering from the lack of repairs.

Bishops themselves, however, were not necessarily moralists, and were in fact urban aristocrats, so baths were still appealing. The promotion and preservation of some baths by bishops could be behind this. It has even been suggested that bishops set up baths for public access, perhaps charging a fee, at least into the sixth century (Miller 2000, 31-2). The scope of these baths was limited, and not as wide-ranging as traditional Roman baths: the aim of these constructions was the urban poor and the needy, not the average citizen (B.

Ward-Perkins 1984, 140). Here is where the rupture with traditional munificence begins.

This rupture, even if slight, marked a point of no return, which would be widened marked in the Islamic period.

The Umayyad state and the Roman aqueducts

Whereas the Roman elites, the Gothic administration, and the Church all had reasons to look back at the Roman past, and saw in aqueducts elements to bridge back to it, the Umayyad emirate did not have to do this. While those aqueducts still functioning offered an indubitable public benefit, the Umayyad emirs had their own agenda, which led them to use and maintain these monuments in a parallel but yet different fashion. Despite the fact that the emirs and caliphs were in charge of preserving public services for the *umma(h)* (the Islamic community), this responsibility was not something which was required from their title or office. Furthermore, the concept of a 'public service' does not seem to have existed in the Islamic world, beyond charitable endowments and pious donations, which were only formally established (like the *wafk*) much later in the Middle Ages. This different attitude requires further explanation, and together with the lack of interest in the Roman past shown by the Umayyads (which offered them no legitimisation), helps explain the relationship between the Umayyad state and the Roman aqueducts, ultimately leading to the end of this public water supply system.

However, the vast majority of the positive evidence available comes from Córdoba, the capital of the Umayyad state, which may not be representative of the overall situation of the post-Visigothic period.

The new legal status of the old water supply

Perhaps one of the most important changes that occurred in this period was the transition from the Visigothic state, with its late Roman infrastructure, to the Umayyad emirate (and, from 929 CE, the Caliphate), which worked on completely different legal and administrative bases. This can be used to explain the full privatisation of Roman long-distance water supplies in the early Umayyad period and the construction of new, private ones.

The two examples we have for privatisation of public supplies come from the suburban aqueducts of Córdoba. During the late eighth century, Abd al-Rahman I cut and diverted the Cercadilla aqueduct, which used to bring water to the Christian complex, in order to supply his own private estate: the almunia of Rusafa (Arjona Castro 2001, 80).[18] Later on, in 967, Al-Hakam II diverted the Western aqueduct in order to bring water to the fountains of the great Mosque (figure 25). For *Reccopolis* and Segovia, the other aqueducts which might have been in use during the early eighth century, we have not enough evidence to discuss their privatisation, but their secondary political role should indicate that they were left untouched.

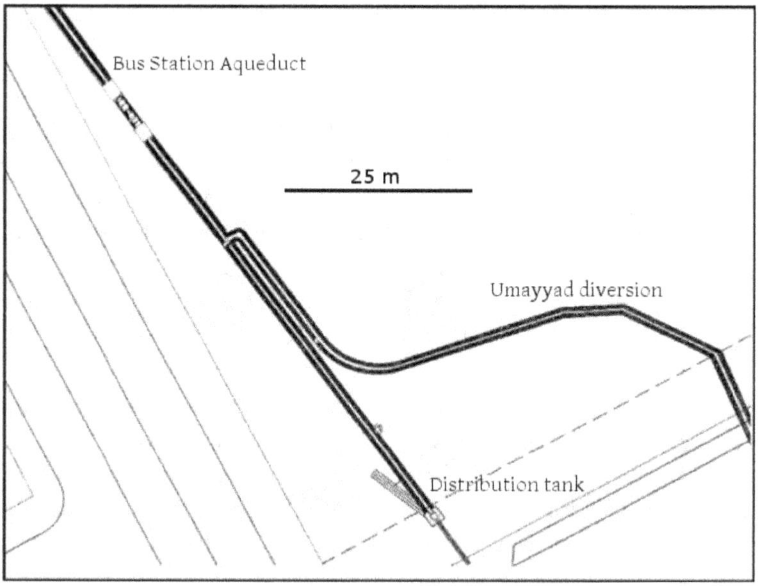

Figure 25: Umayyad diversion of the suburban aqueduct of Córdoba. The water was taken towards the Great Mosque fountains (Pizarro Berengena 2012, fig. 21).

Besides these, we see the construction (or reconstruction) of aqueducts intended only for the private supply of the Umayyad ruling

[18] A practice noticeable already in the East, in Umayyad Jerusalem (Tsuki and Peleg 2017).

elites. The earliest example is in the eighth century, with the construction of the *Qanat amir* (the aqueduct of the emir), which fed water directly to the old Visigothic *alcázar*. During the ninth century we have the reconstruction of the aqueduct of Valencia (figure 26), which brought water across the old forum straight into the palace (Martí and Pascual 2000). During the tenth, and outside Córdoba, the conduit of Valdepuentes was rebuilt to feed the palaces of *Madinat al-Zahra*, the new Umayyad palace city (Ventura Villanueva 2002). Even later on, water supply systems like the twelfth-century aqueduct of Seville, the water-lifting device of the *alcázar* of Toledo, or the monumental water wheel on the Ebro at Zaragoza (both dated to the eleventh century) were built to supply water solely to the urban palaces. These examples show that private monumental, long-distance water supplies had become an established practice in al-Andalus.

This transition of the Roman aqueduct from the public to the private sphere can be partially explained as a result of the new legal status of the public infrastructure. This derives partly from the *sharia* dispositions on water property (the legal principles that justify the privatisation) and partly from the transition of public property after the conquest of 711 (the practicalities of the transition).

From a Classical stance, aqueducts were public in as much as they were property of the council, as per Roman law. Under Islamic law (*sharia*), however, they were not necessarily public. The *sharia* position on water (Montalbano 2008, 690) specifies that naturally occurring springs are communal (even if the most important member of the local community has first rights to access it), but that water sources which are artificially found or channelled are private, of an individual; which seems to exclude the possibility of public ownership. *Sharia* also specifies that small rivers and their deviations are to be communally managed, but this provisions appears to have rural irrigation in mind. Thus, it would seem understandable that the emirs could privatise the aqueducts, as it could be argued both that they were artificial conduits and that in any case the Umayyads had first-serve rights on them. Methodologically, however, this is a problematic interpretation. These legal dispositions of the *sharia* had not been firmly established in the early Umayyad period, and would only become normative in the ninth and, especially, the tenth century.

They were established *post factum* as a way of legitimising what had already been done. The question is how far they reflect pre-existing norms, and how much they were means of legalising opportunistic usurpation.

Figure 26: The Umayyad conduit of La Almoina in Valencia, built diverting the old Roman conduit to the governor's palace across the old forum (photograph courtesy of the SIAM, Valencia).

Focusing on the actual practicalities of the eighth and ninth centuries, what can be seen is a transfer of property after the 711 invasion. The aqueducts were, as we have seen, public property controlled by city councils. During the Umayyad conquest most cities surrendered through treaties known as *sulh*, by which the locals were allowed to preserve their possessions and status in exchange for extra taxes (Chalmeta Gendrón 1994, 213-20).[19] This would assume that in the first decades of Damascus control municipal affairs were left to their own. During the process of state formation of the independent Umayyad emirate, the narratives and accounts of the conquest were modified (Clarke 2012), so to justify more conquests by force that would render cities unprotected by the *sulh*, and thus would have the public properties transferred to *umma* (communal) ownership (Manzano Moreno 2006, 39-40). The *umma* was the community of believers and the emirs and caliphs were, by definition, their leaders both in a religious and political sense (Crone 2005 [2004], 3, Santillana 1926, 2). This means that as leaders they had a degree of control over communal properties. Perhaps in order to further justify the overall privatisation, it could be the case that the aqueducts were considered part of the ruler's share after the conquest (mirroring the transfer of other Visigothic royal properties), and that the town council (in whatever form it still existed, if at all) could or did not contest this (Santillana 1926, 134-6). But again, all our texts date to a later period, and are part of an Umayyad effort to justify retroactively their rule and decisions. The nature of the sources blurs our approach to the actual events and transformations.

Despite these privatisations and the lack of investment in large-scale public water supply in the Roman way, the Umayyads were responsible for some water-related munificence. After all, the fountains of the Great Mosque may not have been a main watering point for the local inhabitants (their ritual role was surely paramount), but it was a pious donation of water. Similarly, the creation of a fountain outside the *alcázar* by Abd al-Rahman III with the overflow of the *qanat amir* (*Chronicle of al-Nasir*, 28), was a way of giving to the

[19] Only one treaty has been preserved, that of duke Theudimer.

people part of the water which the palace had not consumed (thus sticking to the principle that the Umayyads had first-serve on that source, but were still expected to share it). However, these actions were done as pious goodwill and charity, and not out of civic duty as it was the case in the Roman period.

While there are many gaps in this legal proposal, especially considering the role or existence of city councils, and the relevance of later Islamic law for the early period, it can be said with some certainty that the Umayyad attitude towards the Roman water supply shows a clear rupture with the late antique model. The lack of a clear definition of public or municipal property and the new Islamic precepts on water law favoured this transfer of ownership to the Umayyad elite. Water munificence as it had developed in the Roman and late antique periods disappeared fulfilled no political or legal role in al-Andalus.

Looking back, looking East: Syria and al-Andalus

The development in al-Andalus of a new Umayyad state in two phases, the Emirate (754-929) and the Caliphate (929-1031), is linked to two periods of urban renewal, as the emerging power developed a new public architecture to make itself visible. This is very similar to the process which the Suevic and Visigothic kingdoms went through earlier, in the late sixth century. However, and different from what we have seen for the Visigothic period, where the Roman past was an element of legitimisation, the Umayyads did not need to look back to the Romans or their monuments to justify their rule. In fact, the Umayyads looked back at their own past as Caliphs in Damascus (Clarke 2012, 40-1, Fierro 2007).

Even if from a strategic point of view the Umayyads favoured as their principal nuclei those centres which had been major towns during the Visigothic and Roman periods (Zaragoza, Mérida, Toledo, Córdoba and Seville), they translated the infrastructure and administration from the East, from the Umayyad and Abbasid caliphates (Chalmeta Gendrón 1994, 26-8, Lévi-Provençal 1957, Souto Lasala 2001, 29). For them, the Roman towns and their monuments were not something to look at for inspiration or prestige.

Figure 27: The Great Mosque of Córdoba.

The Great Mosque of Córdoba is the best example of this. Built late in the eighth century by Abd al-Rahman I, it became not only the main mosque of al-Andalus, but also a dynastic monument which all the Umayyads after him enlarged or repaired (figure 27). The reuse of Roman and Visigothic columns and capitals was a way of indicating a transition between the 'old' and the 'new' regime, and an appropriation of that past giving it a complete new meaning (Cressier 2001, 319-23). The design of the mosque itself, despite some claims that the superimposed rows of arches imitate the aqueduct of Los Milagros in Mérida,[20] indicates a Syrian tradition. This is clear not only in the arches (Syrian horse-shoe arches, and superimposed arcades as in the Mosque of Damascus), but also in the roofing technique (which imitates the al-Aqsa mosque at Jerusalem), the proportion of the aisles, and the use of alternating colours of masonry: the so-called *ablaq* technique (Clarke 2012, 40-1, Creswell 1940, 156-7, Souto Lasala 2001, 30). All these architectural elements have Roman precedents, but the way in which they were used in al-Andalus was a fully Islamic one, which the Umayyads may have taken from Roman Syria, but which, by the time they had reached Spain, had lost any symbolic Roman meaning. The mosque had become such an important dynastic monument that when it was found out that the *qibla* wall was facing some 50º off true *qibla* (preserving the alignment of the Roman street grid), Hisham II decided to preserve the original orientation of Abd al-Rahman I (Manzano Moreno 2006, 315-8).

With these clear points of reference to legitimise their position (their own past as caliphs in Damascus, the Abbasids, and during the Caliphate of Córdoba, Abd al-Rahman I himself), the Roman past seemed little else than a background noise. In fact, in Arabic sources there is a large-scale ignoring and discrediting of the Latin past (Elices Ocón 2017, 274-7). Aqueducts had proven very useful, and, as it has been shown, they had been in fact preserved to an extent, but then these monuments had no symbolic importance whatsoever. Aqueducts were not linked, in Umayyad context, to civic duty (Elices

[20] The arches of Mérida do not alternate colours, and the aqueduct itself has three rows of arches. This claim is completely preposterous and dangerous, as it links to revisionist and islamophobic approaches that deny the Islamic invasion of Spain (García Sanjuan 2016).

Ocón 2017, 407). Those aqueducts which were not useful for the Umayyad agenda or were beyond repair were ignored. Those which were still standing but not flowing were left mostly unrepaired, and when they were restored (as in Seville and Valencia, perhaps in Mérida) they were used privately for the local rulers, as mentioned above. Under this perspective, it is difficult to defend the 'submission of the Arabs to Rome through its utilitarian works' proposed by Basilio Pavón (1988, 418).

The importance of Roman aqueducts in Umayyad urbanism and town planning was very limited. It is certain that the means and knowledge to preserve and repair aqueducts were available in the Umayyad period, but with hardly any political interest in actively doing so, aqueducts had become purely functional structures. Furthermore, whereas bridges could rank together with aqueducts in terms of urban importance, the former (especially that of Córdoba) were thoroughly repaired, as they were much more useful in Umayyad schemes and ideas of urbanism than aqueducts (Zozaya 2001, 55).

Umayyad urbanism and water supply systems up to the end of the Caliphate

Overall, the development of the Umayyad state(s) was linked to the emergence of its own new public architecture and its new urbanism. This new urbanism broke with the previous (late Roman and Visigothic) trajectories of urban development and, either because of the long period of disrepair of the infrastructure or the attitude of the Umayyads towards the old aqueducts, Roman water supply systems did not play a major role in these new urban developments. In fact, public cisterns (*jubbs* or *aljibes* in modern Spanish) seem to have been far more common, together with professional water-carriers who delivered water from these sources door-to-door (Glick and Kirchner 2000, 303, Navarro Palazón and Jiménez Castillo 1995).

The general consensus is that throughout the Umayyad period towns grew and developed around mosques and the main palaces, which were mostly built in the core of the previously existing towns, and were increasingly fortified. The street hierarchy was preserved, although the hierarchy was not reflected in width or straightness (A. Almagro Gorbea 1987, Canto 2001, 13, Gutiérrez Lloret 2012, 197, Lévi-Provençal 1957, 198-200). Therefore, there is a continuity of the principal foci of urban development in the main towns of the al-Andalus. What seems clear from the evolution of walled enclosures is that towns in the Umayyad period (especially after the ninth century) grew and expanded – not only the main settlement, but also its immediate suburbs. This growth and these developments took place without a regular water supply in the Roman fashion. In fact, and looking back at the patterns of urban settlement outlined for the Visigothic period above, it is not surprising to see that the location of the Umayyad urban centres coincides with those of the Visigothic period, and not necessarily with those of the Roman one.

Figure 28: The Alcazaba of Mérida (photograph by Isaac Sastre).

The clearest example of this would be Córdoba, as the mosque and the *alcázar* are built on top of the Visigothic *palatium*, located by the bridge and where the water table is available (Ruiz Bueno 2017). The same happened in Zaragoza, in Tarragona with the upper town, and in Valencia with the *alcázar* built on the episcopal area. Even perhaps in Barcelona, where it has been suggested by the team of Luis Caballero that the 'episcopal palace' could in fact be an early Umayyad building, if compared in size and plan to the Umayyad palaces excavated at Morería and the temple of Diana in Mérida (Martínez Jiménez, Sastre de Diego and Tejerizo García 2018, 279). In these examples there was continuity in the role and importance of given areas from the Visigothic into the Umayyad periods, and only in Tarragona (and maybe too in Valencia) these coincide with the centres of Roman urban power.[21] This is not only an indicator of the high degree of continuity between the Visigothic and Islamic periods, which accounts for the early settlement and administrative patterns of the Umayyads (Gutiérrez Lloret 2012, 197, Manzano Moreno 2006, 248-60), but also of the importance artificial water supply systems had in modifying the urban layout. Only in very limited examples did the Umayyads develop a brand new area during the ninth century, consciously trying to break the links with the Visigothic past: in Mérida and Toledo the Umayyad fortifications included a new fortress in a strategically located area to control the unruly populations of these towns, as their strong local aristocracies were responsible for continued rebellions. In the former, the *alcazaba* was built after the destruction of the walls on top of the Gate of the Bridge (figure 28), which had been the symbol of the Roman colony since its foundation (Alba Calzado 2001, Creswell 1940, 198); in the latter, the *alcázar* was built on the highest point of the granitic dome of the old Roman town after the destruction of the suburban *praetorium* (Izquierdo Benito 2009). Curiously enough, in both instances, the deliberate location of the fortifications required the construction of new water supply systems, as they were located in areas without an obvious one. In Toledo, a set of water-lifting wheels was eventually installed. In Mérida a new *aljibe* or filtration cistern was built in the central area of the fortress, reusing Visigothic pilasters in its decoration (showing the power of the new rulers over the past) and conveniently orientated towards Mecca, so a mosque could be installed on top (figures 29, 30) (Cressier 2001, 319, Creswell 1940, 202-3).

---

[21] The case of Valencia is perhaps debatable, as the *alcázar* is already outside the old forum area, adjacent to the old episcopal complex; but as this is now underneath the Gothic cathedral, it is very doubtful that any excavations will be carried out there.

Figure 29: Aljibe of the Alcazaba.

In terms of domestic architecture, communal and private wells and cisterns were abundant in residential areas. Fountains were mostly found in mosques, and even these seem to have been linked to cisterns and wells. There were no more links with aqueduct water supplies, and towns fell back to pre-Roman patterns without much problem. In terms of sanitation, however, this caused some alterations: while conduits to evacuate domestic sewage were not common in al-Andalus, cesspits are far more abundant. The only concerns Islamic law-givers seem to have had, which may be a reflection of the

Figure 30: Decorated pilasters from a Visigothic building reused as part of the aljibe.

simple drainage system, was that sewage and rain water should not flow on the same drains. That may be why open-air gutters to drain rainwater seem to have been the basic and most common type of sewer – and these did not rely on constantly-flowing water from an aqueduct (Rėklaitytė 2012, Vidal Castro 2000).[22]

Considering the previous points, it is also noticeable in new urban foundations that there was a lack of overall public water supply systems. In *Madinat al-Zahra* the aqueduct was used to feed the palatine complex, despite the fact that the residential areas were densely occupied (Acién Almansa 1987, 12-8). The town of Jaén, refounded in the Umayyad period, obtained its water from springs, which had been monumentalised in the Roman period, but the distribution conduits had long been abandoned, and were not restored in the Islamic period (Salvatierra Cuenca, Serrano Peña and Pérez Martínez 1998). Similarly, Anjar in Lebanon was a new Umayyad urban foundation, settled in an area where the water table was easily accessible, but without a public infrastructure to distribute or administer it (Carver 1996, 189-90, Haase 1996, 167).

Lastly, Islamic baths or *hammams* were very numerous in al-Andalus. Three hundred are mentioned in the sources in Córdoba alone, although archaeologically we only know a handful (Clapés Salmoral 2013). There are still standing baths in Palma de Mallorca, Granada, Ronda, Málaga, and many other places. In recent years, a set of baths in Toledo has been identified, next to the mosque of Bab al-Mardum (now chapel of the Holy Cross). These baths all functioned with small quantities of water, as they lacked the large pools which characterised Roman baths. Cisterns or water wheels were the main water supply system for these baths, although the *hammam* of La Almoina in Valencia might have been fed by the aqueduct, and the baths of Bab al-Mardum in Toledo were linked to an underground water filtration gallery (Passini 2010).

---

[22] There are a few notable exceptions, as Córdoba, Zaragoza or Mérida, where some Roman sewers were preserved or repaired, or Lérida and Murcia, two new urban foundations (Remolà Vallverdú and Acero Pérez 2011).

An aqueduct-less society

Despite the fact that the techniques and the knowledge to build and maintain aqueducts were available in the Islamic world, these were never as common in the towns of al-Andalus as they were in the Roman period (Hill 1993, 183-5). Water was still an urban need, and judges especially were very concerned with the availability of water in urban contexts, but this was rarely supplied by Roman aqueducts (Navarro Palazón and Jiménez Castillo 1995, 402). Old Roman aqueducts had been through a long period of disrepair after the fourth and especially the fifth century, so in many instances it would have been necessary to re-establish the aqueduct supply at a high cost, which the new Islamic authorities could not afford, and which they did not care that much about. There was probably also no popular urge to put them back in service.

This fits the general trends of urban development in the early Islamic world, where the only true necessary and common elements in towns were fortifications, the mosque, and the palace (Kennedy 1998). All of these are known to have been quickly built in Spanish towns soon after the conquest (Zozaya 2001, 54-5). *Hammams* and *suqs* or market areas would later develop through private initiative. The presence of other public infrastructure was mostly up to the interest of the ruling elite, as the lack of a municipal authority that encompassed most of the urban aristocrats prevented the construction of large public buildings, for which there was no real demand as it existed in the Roman period (Salvatierra Cuenca, Serrano Peña and Pérez Martínez 1998). This return to pre-Roman modes of water supply should not be seen as a step back in urban development; after all, most Islamic towns grew far beyond the limits of their old Roman wall enclosures.

Later in the Medieval period, new long-distance water supply systems were built in Islamic towns in Spain,[23] as in Palma de Mallorca (Riera Frau 1993), Madrid (Retuerce Velasco 2004, 101-4,

[23] Usually called 'qanats' in Spanish publications, using the old Arabic word, although they did not function in the same way modern scholars understand it, where a *qanat* is understood as a *foggara* (Hodge 2002).

Valdés Fernández 1992, 158-61) or Huelva (Pavón Maldonado 1990, 205). In other instances, rural irrigation conduits were diverted on a regular basis to fill domestic and public cisterns and baths. But these two models are usually much later in date (late tenth century at the earliest) and cannot be linked to Roman attitudes to public water supply.

CONCLUSIONS

Aqueducts evolved from being ever-present monuments that each Roman city worthy of the name wanted to have, to an established part of the urban infrastructure, and, finally, a derelict reminder of a past era. This was a general trend all over the Empire, but perhaps it is in the West that these changes were most notable. The examples put together for the Iberian Peninsula reflect these trends, while casting some light on the actual evidence. The most important conclusions that have been put forward are three. One, that aqueducts had a surprising degree of continuity into the early Middle Ages. Two, that this degree of continuity was the result of an equilibrium between the inherent usefulness and legitimising Roman appeal of aqueducts and the incapability (due to lack of resources, knowledge or skill) to keep them working. And three, that even if originally aqueducts had not been essential parts of towns, by this period they had an immense influence on urban layouts, so their abandonment was one key factor in the shaping of urban settlements.

A SURPRISING DEGREE OF CONTINUITY?

One of the big questions that this book has tackled is the degree of continuity aqueducts had, a question that had not previously been addressed or answered, and was most usually just ignored. It is possible now to put forward figures on this continuity (figure 6), and to claim that of the 69 aqueducts of the Iberian Peninsula, one in four (24.6%) were still in use by 400 CE and into the fifth century (Almuñécar, Barcelona, *Conimbriga*, Córdoba, Lisbon, Lugo, Mérida, Segovia, Seville, Tarragona, Valencia and Zaragoza). In some cases, as mentioned, the dating is quite secure, as in Mérida, Seville or Córdoba, whereas in others the certainty is lesser (Almuñécar, Valencia), but they form a cohesive group of datable aqueducts. The number of functioning aqueducts declines steadily, but gradually, through the fifth and sixth centuries, with a steep fall into the seventh. This means that by the Umayyad invasion only four of the old Roman aqueducts are in use, plus two other doubtful ones, and the new sixth-century aqueduct of *Reccopolis*.

Unsurprisingly, those aqueducts that show continuity are linked to the most important towns of the Peninsula, those in which a powerful elite that could take care (or try to maintain) their water supply system was still present.

It would be very difficult to compare Spain to other regions, such as Gaul or Italy so as to assess if this ratio is a good indicator for the post-Roman West – although such a study would be extremely interesting. This is not only because there is no comparable corpus of data for either of these regions, but also because the local dynamics that ruled the years between 400 and 700 are very different. Italy had a much larger number of urban aqueducts and, for the fifth and early sixth centuries, there was a much greater degree of Roman continuity in all aspects. But this all changed after the devastating wars of the mid-sixth century. As for Gaul, it is similar to Spain as it had a Mediterranean core which long preserved *Romanitas* (in fact, most of Mediterranean Gaul was ruled by the Visigoths) and a wide strip of northern and eastern territories which became increasingly deurbanised. However, the geomorphology and climate of Gaul do not really make aqueducts essential parts of urban development, as water was much more easily accessed than in Spain.

One last key thing that can be inferred from the chronology and the data presented in this book is that the approach to aqueducts was very different in the Roman period than its later Umayyad (and subsequent) equivalent. Aqueducts had been key in early Roman urban culture, and were still present in the late Roman period, lingering through Late Antiquity, but after a period of abandonment, the reuse and reconstructions of the Islamic period did not follow the path marked by the Romans.

THE LONG SHADOW OF THE PAST

From the written sources of late antique Gaul and Italy we get the impression that preserving the Roman past was something very important for the local elites and the late imperial administration. The old Roman monuments represented a legitimising link, which they wanted to take over. In *Hispania* the lack of written sources prevents us from making strong claims on this aspect, but the attitude towards aqueducts may shed some light on it. It is mostly the urban elites in the important towns listed above that seemed to have enjoyed aqueduct supply, and in theory the ones who were responsible for its upkeep. Bishops seem to have also been involved up to a certain extent in these affairs, as they represented the new urban elites, as the possible episcopal baths of Barcelona or the episcopal buildings at the forums of Valencia and Tarragona suggest. The use given to the public water supply system had, however, started to shift into a more private sphere, and although some public fountains are still present and in use (like Tarragona, Valencia), the aqueduct supply seems to have been preserved primarily as a private elite supply. This itself is not that different from the early Roman period, but the lack of public baths made aqueduct supply much more restricted, and therefore, much more elitist.

Despite this, aqueducts were still very useful and powerful symbols of *Romanitas*, so it is not surprising to see that a new royal urban foundation such as *Reccopolis* was equipped with a new aqueduct. This major construction has to be put in context, as an aqueduct was part of the 'ideal town', it was one of the things that any important capital (especially in the context of Rome, Ravenna or Constantinople) should have, together with a church, a palace, and walls. The lack of any other major royal constructions in Visigothic Spain (beyond wall repairs) may account for why there were no other aqueducts built by the monarchy, but it does not explain why the monarchy was not responsible for the repair or maintenance of other urban aqueducts. Perhaps, and following late antique traditions, municipal infrastructure was increasingly left aside by the patronage of the central administration.

This approach towards aqueducts from the local elites and the local Church (or from the central administration) came to an end with the fall of the Visigothic kingdom, as in the Umayyad period and beyond aqueducts became completely privatised. This privatisation included the rebuilding of aqueducts to supply private estates and palaces of the Umayyad ruler, even if this involved cutting the public supply. During the tenth century the Umayyad caliphs used aqueducts to supply water at the mosque of Córdoba, as a sign of the caliph's piety and goodwill. But this was a gift to the mosque (which was an Umayyad dynastic monument), rather than a large-scale fountain distribution commissioned by the city council which had characterised the Roman period.

THE DECLINE AND FALL OF URBAN AQUEDUCTS

The demise and ultimate abandonment and collapse of aqueduct is the main questions addressed in this work, and various answers have been put forward. The transformations of the urban administration in the fourth century caused many of the transformations that are noticeable in the fifth century, including the lack of investment in aqueduct maintenance. The lack of municipal funds and the lack of political incentives during the late fourth and fifth centuries are perhaps the main cause for aqueduct disrepair. Only basic maintenance was carried out in these centuries, if at all, so it is possible that as the demand for specialised and skilled engineers during this period declined, by the period of urban renewal of the late sixth century there were hardly any (if any at all) engineers who could repair major damage on aqueducts. This meant that these structures, which were mostly over four hundred years old, were impossible to repair and, wherever they tried to repair aqueducts, they failed (as can perhaps be seen in Mérida). In this context, the aqueduct of *Reccopolis* is even more striking, as it is built *ex novo* in a context in which this did not seem possible.

This abandonment of the water supply infrastructure did not, however, imply that towns were left without water. Some areas of towns lacked running water, but water was still available (as it had been in the previous periods) from wells and cisterns. In some cases, these become more numerous as the aqueducts cease to function, and in some other cases, due to the lack of aqueducts, large public cisterns had to be built (as in the new urban foundation of El Tolmo).

Aqueducts had not originally been an urban need, but rather a way of showing allegiance to Rome and of presenting the locals as fully Romanised (the famous Monty Python's '–And what have they [the Romans] ever given us in return?!'[1]). But as time passed and the benefits of piped water became more evident, urban populations grew more and more dependent on aqueduct supply, up to the point that in the late antique and post-Roman period the end of aqueducts prompted changes in urban layouts. We have seen through various examples of the main towns of the Peninsula that townscapes and settlement location changed through this period, and even if there may be other reasons for this (the relocation of the centres of power, new economic foci, etc.), water availability was certainly one of the most important. This is especially true in upper towns which were once fed by aqueducts, which may have become less welcoming without public fountains or domestic piped water.

This statement must be kept within perspective, as towns flourished before and after aqueducts were available: towns in the Islamic period and in the later Middle Ages grew beyond the Roman or late antique enclosures. Only in a very few cases were aqueducts a real urban need. The important thing to keep in mind is that in the post-Roman period town dwellers had to adapt to a situation they were not prepared for, in which water ceased to be a commodity and had to be acquired from wells or cisterns.

FURTHER AFIELD?

In conclusion, this book has provided not only innovative results but also a clear general overview of a few chosen late antique cities in *Hispania*. And even if this may be a useful contribution, there are still things that can be investigated and blurry areas that need further research. The most obvious thing, as already mentioned above, would be a similar study to assess the evidence for Gaul, Italy and North Africa: a general overview of the end of aqueducts in the west

[1] It should be noted that the first thing mentioned in the film out of the long list of Romans did actually *do* for the Jews in Jerusalem was, in fact, the aqueduct.

would be very useful to improve our understanding of the evolution of towns in the post imperial period. Similarly, with a larger body of comparanda, it will be possible to improve the results and refine the models proposed in this book. Jordan Pickett is currently (expected 2021, Oxford University Press) preparing a volume on aqueducts in Byzantium which may make these issues clearer.

Further research in absolute dating of aqueducts would also be extremely useful. ESR (electron spin resonance) and OSL (optical stimulated luminescence) dating can date the crystallisation of carbonate, and therefore, of sinter. With proper sampling *in situ*, it is possible to micro-sample crystals from the bottom and the top of the sinter concretions, and together with a Bayesian model, it could give thus dates for the first and last flow of water in the conduit. This type of sampling and dating has not yet been applied to aqueducts, but it has been used to date enamel formation in Palaeolithic teeth (Duval, Grün, et al. 2009, M. Duval, C. Falguères, et al. 2012). Modern developments of micro-sampling would further allow us to date the formation of crystals from the top and the bottom of the sinter crust, giving us an absolute date between the last time sinter was removed and the date in which water ceased to flow.

From a more local point of view, a closer analysis of water-related material culture in post-Roman urban contexts (such as the analysis and volume of water vessels and jugs, or plotting well distribution in a single site) could provide a much more detailed account of the dairy uses of water. This could be compared across various sites of Roman, late antique and Islamic date. Much is yet to be learned from the many boxes of fine and common-wares excavated in recent years and stored in museums, which have not yet been studied.

Lastly, in the period between me finishing my thesis (2013) and its final publication (2019) nine more aqueducts have been identified or published in the Iberian Peninsula, which means that more raw data may still emerge.

***

Overall, aqueducts in the towns of post-Roman Spain and Portugal were mostly standing, but not necessarily in use, as silent reminders of a previous era. A considerable number of them were still functioning, especially in the most vibrant towns where urban elites still had enough resources and power, which is one of the most important conclusions for this book. But as generations passed and Roman rule became more of old people's memories, the local capability of preserving and repairing their water supply systems diminished. It is doubtful that aqueducts ceased to function overnight, and may have taken various years for an aqueduct to finally cease to function – but in that time the inhabitants may well have come to realise that they were living under new circumstances. The spell of aqueducts was so appealing that the Visigothic king's new town was equipped with one, and again – one just can wonder about what the reaction of those newly settled inhabitants was when they saw it. It would have to be the newly-arrived Umayyads, with their different attitude towards the Roman past, and their inherited and preserved knowledge of hydraulic engineering that brought many aqueducts back to use – but by then the population did not benefit from them, and probably they did not expect to either. But still, and as many of tourists these days, they would have looked at the centuries-old aqueducts which were still standing and wondered about these most impressive monuments.

CATALOGUE: FUNCTIONING AQUEDUCTS IN THE LATE ANTIQUE IBERIAN PENINSULA

ALMUÑÉCAR

Almuñécar (the ancient *Sexi*) was a Phoenician colony in southern Spain, which became a *municipium* in the early Empire, and this promotion prompted a period of urban development which included the construction of an aqueduct and a set of baths linked to it (Sánchez López and Moreno Pérez 2012). The city itself was built on top of a hill, overlooking a harbour, and it seems to have been primarily a centre of *garum* production, like many other coastal towns of this region.

Very little is known of the archaeology of Almuñécar in Late Antiquity, but it seems that by the time of the Byzantine invasion (550s) the city was largely abandoned, including the large *garum* factories at the site of El Majuelo. The city had already ceased to mint coin in the fifth century, and even if it might have still preserved some minor importance as a harbour, by the seventh and eighth centuries the site may well have been deserted, as no material is known from the site or its environs datable to that period (Gómez Becerra 1995).

The site still functioned as a harbour during this period, even if the archaeology indicates that it was a rather small one, because in 750 CE Abd al-Rahman I landed here after fleeing from the Abbasid Revolution there. The site shows signs of commercial and urban activity again in the ninth century.

The remains and course of the aqueduct

The aqueduct of Almuñécar (Sánchez López and Moreno Pérez 2012) is fairly well known all along its 7km, and its impressive remains are still standing and, in some sections, still carrying water for the neighbouring fields (figure 31). It draws water from an aquifer up in the mountains around Almuñécar (the site of La Angostura), and by means of a partially subterranean conduit takes the water along the valley of the Verde river, visible at Barranco de Antequera and Barranco del Olivillo. In this part, the aqueduct goes over the Torrecuevas creek on an impressive seventeen-arch structure. A few metres away, the aqueduct goes into a tunnel to the valley of the Seco river, on the other side of the mountain. There, the creeks and ravines are deeper (figure 32), and up to three bridges were built (numbered aqueducts I, II and III), which vary between 43m with four arches (I) and 72m with nine arches (III).

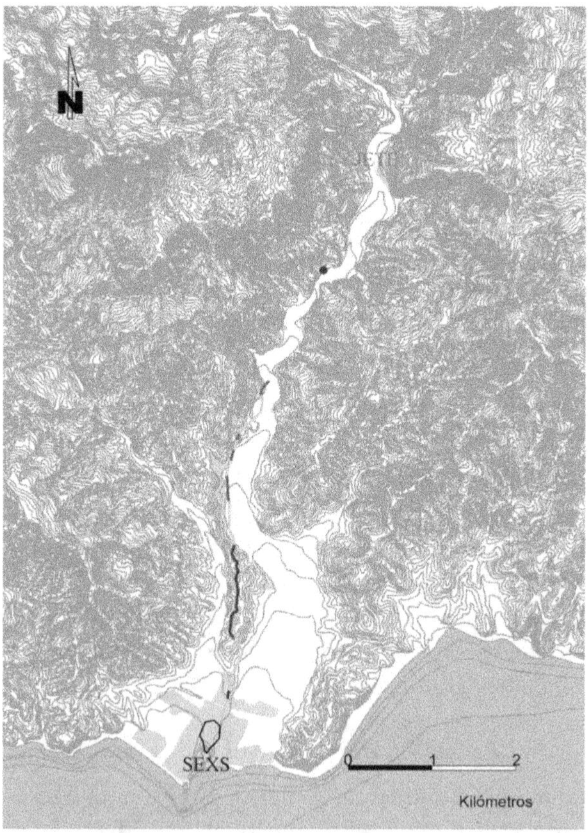

Figure 31: Map showing the course of the aqueduct of Almuñécar.

Figure 32: Arcaded section of the aqueduct (photograph by Elena Sánchez).

After aqueduct III, the conduit reaches a settling tank which is the head of an inverted siphon that takes the water up to the hill where Almuñécar is, crossing the valley in between. The *venter* (low point) of the siphon has been located recently; it ran on clay pipes over the deck of an arched structure (figure 33).[1] The final tank of the aqueduct was for a long time believed to have been the 'cave of the seven palaces', a Roman vaulted structure, but this lacked any water-tight lining (according to the recent excavations), and it seems more plausible that the current bell tower of the Encarnación church (which is a reused Roman tower) was the aqueduct's final destination. The tower would allow the de-pressurisation of the water, which would match the description given by al-Idrisi in the twelfth century, who mentions:

---

[1] Such *venter* was predicted by Fernández Casado (1949), and documented in the 1990s (Bestué Cardiel and González Tascón 2006).

> '... [in Almuñécar] there is a square construction, similar to a column, wide in its base, narrow in its top. On two of its sides there are carvings [imprints of pipes?] that merge together on the top. At the angle formed by the side there is a sunken vat in the floor, which received the waters brought from a mile away by an aqueduct of many arches built in hard stone [at this distance, it must have been the venter of the siphon]. The learned men of Almuñécar say that in former times the water went up to the top of the obelisk and went down the opposite side, towards a small mill...'[2]

Figure 33: Venter of the aqueduct's siphon (photograph by Elena Sánchez).

A branch of this aqueduct may have been identified at the El Majuelo *garum* factory, at the foot of the hill of Almuñécar, which brought the water needed to clean the fish and vats. Although the connection is not confirmed, the presence of a vaulted conduit lined in *opus signinum* could only point towards an aqueduct; if it was not coming from the main aqueduct of Almuñécar, it must have been a new conduit that brought water from elsewhere (Sánchez López, Pérez Marrero, et al. 2011).

Chronology and dating evidence

Various Arabic sources (al-Himyari, ibn al-Khatib) mention the aqueduct of Almuñécar or its terminal deposit, copying the text of Idrisi quoted above. But all of them mention it as a ruin which no longer functioned, although it was obvious to them that it was a water con-

[2] Quoted in Fernández Casado (2008 [1972], 205). My translation of his Spanish.

duit. Al-Idrisi, the earliest of these writers, wrote in the early twelfth, giving us a rather vague and unhelpful *terminus ante quem*.

For an earlier *ante quem* we have to look at the lack of datable pottery for the seventh century from the town itself, which would indicate a decrease in the activity of the site parallel to other similar nuclei in the Byzantine province, like Málaga, *Carteia* or *Traducta* (Martínez Jiménez and Moreno Narganes 2015). Presumably by then the aqueduct had already ceased to function.

The only other dating evidence is the abandonment of the fish factories at El Majuelo, dated to the late fourth or early fifth, although the pottery has been recently reassessed as late fifth (Gómez Becerra 1995, 185, Lagóstena Barrios 2001, 154), giving a possible *t.a.q.* for the end of the aqueduct. The end of the fish factories of Almuñécar could in fact be directly related to the end of the conduit, although this is not terribly useful dating evidence. The baths located at the *venter* of the siphon were very badly damaged by ploughing, and have not provided any dating material for its late and abandonment phases.

In the best of cases, the aqueduct may have been working into the fifth century, and most certainly not into the sixth, although there is no hard evidence.

BARCELONA

Barcelona (*Barcino*) was during Late Antiquity one of the most important towns of the north-eastern Iberian Peninsula. The old Roman colony had been founded on two hills between two small rivers overlooking the Mediterranean coast, and it was back then a secondary regional town within the area of influence of Tarragona, the provincial capital. Its political importance grew during Late Antiquity, and during the Visigothic period it replaced Tarragona as the regional capital.

The increasing political importance of late antique Barcino

Barcelona began to develop as a political centre during the early fifth century (Ripoll López 2002, 34), when it became the seat of power of Athaulf and Galla Placidia (410-15), probably because of its location close to Gaul, its good harbour, and its walls reinforced during the course of the fourth century (Pallarés Salvador 1969, Puig Verdaguer and Rodá de Llanza 2007). It became part of the Visigothic kingdom in the 470s (Perich Roca 2014), and it seems that there was a palace-like structure used by the Goths as a royal seat there (*Chr. Caesaraug.*, 510; *Hist. Goth.*, 38). However, it was because of the presence of a bishop that Barcelona became the key city in the province of *Tarraconensis*.

The city had been an episcopal see at least since the fourth century, and it was both a Catholic and an Arian bishopric until the third Council of Toledo in 589 CE. In the year 592, King Reccared invested the bishop of Barcelona with the power to collect the taxes in the territory of three other adjacent bishoprics, including the metropolitan see of Tarragona, in a document called *De fisco Barcinonensi.*[3] The power of the bishop was thus increased, which seems to be reflected in an enlarged episcopal complex, located in the north-eastern quarter of the town.

The episcopal complex of Barcelona (figure 34) is a very good example of well-researched urban archaeology. The excavations below the Plaza del Rey, the cathedral, and the City History Museum have revealed a large site which was most probably the political centre of Barcelona. This has been identified as the episcopal complex, which can be dated to the fifth century, with an important expansion in the sixth. The complex and the cathedral were originally (4th c.) located on a workshop area which included a *fullonica* and *tinctoria*, and a fish sauce factory (Beltrán de Heredia 2002b, 2002a). The *episcopium* back then consisted of a large residence constructed over a previous *domus*, the cathedral basilica itself, and an audience hall, all of which were contained into a single street-block (Bonnet and Beltrán de Heredia 2002). At a later date, probably during the late sixth century (perhaps related to the powers granted to the bishop by the *De fisco Barcinonensi*), the whole area underwent a major reconfiguration: the buildings were finally abandoned and substituted by new structures belonging to the episcopal complex: the cathedral was enlarged, the baptistery was remodelled, a new cross-shape building (identified as a church) was erected, and a new U-shaped building

[3] Discussed by Chris Wickham (2005, 96-7). Cf. Fernández 2006.

(suggested to have been the count's palace) was built. In order to build this large complex several transverse streets were blocked and included inside the complex. A new bath complex was built in the sixth century too, as will be set out below.

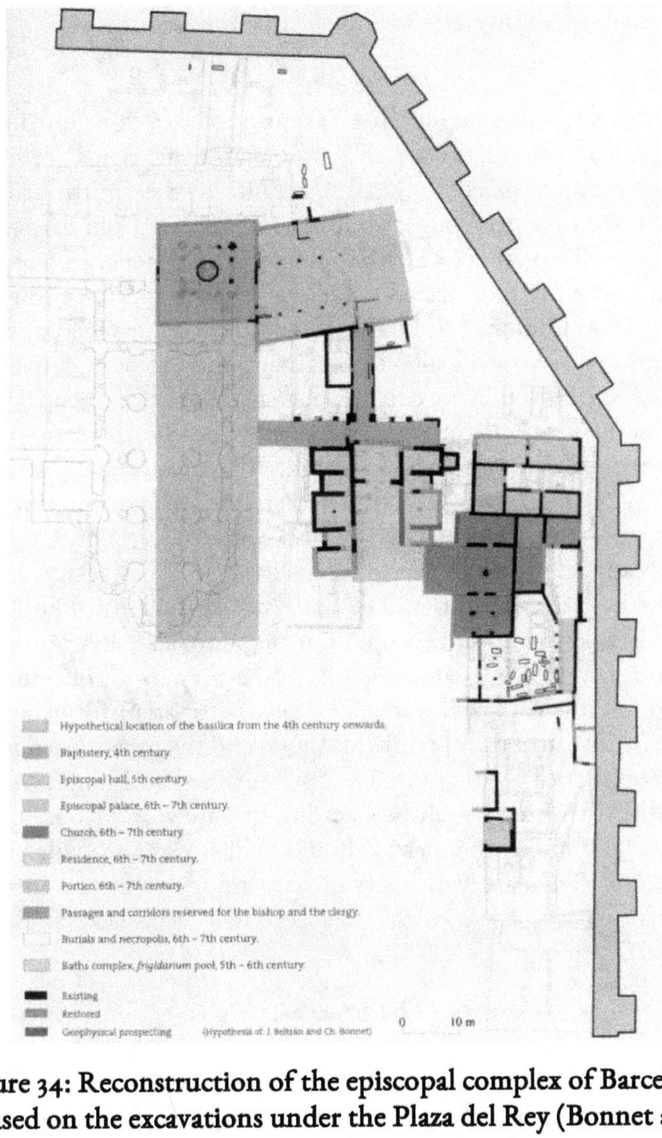

Figure 34: Reconstruction of the episcopal complex of Barcelona, based on the excavations under the Plaza del Rey (Bonnet and Beltrán de Heredia 2002, fig. 9).

The rest of the town remains largely unknown for this period, but it seems, from the evidence of extensive reuse of materials in the episcopal complex, that whatever traditional public structures still survived were pulled down in order to obtain new building material (Beltrán de Heredia 2002c). Outside the walled town, a new sixth-century suburb developed towards the north-west.

Late antique baths

Of the few structures known from late antique Barcelona outside the episcopal complex, baths are the most abundant. Some new large baths were constructed during the course of the fourth century, which might have been public, but it is not known when they ceased to function. Two other late baths belonging to aristocratic *domūs* are known: those of Bisbe Caçador (belonging to a fourth-century *domus*) and Pati Llimona (5th c.). Both seem to have continued until the sixth century at least, parallel to the abandonment or re-conversion of the *domus*, although the dates of these baths are not certain.

The *balneum* of the *domus* of Bisbe Caçador is a large set of baths, with five rooms which include an *apoditerium* [changing room], two *frigidaria* [cold rooms], a *tepidarium* [warm room] and a *caldarium* [hot room], paved in *opus sectile* (García-Entero 2005, 207). It was built on top of the *intervallum*, so it has been dated to the fourth or fifth centuries. The bath furnace was shut by the end of the fifth century, but this does not prove that the bath ceased to function; it rather indicates that the baths may have continued to function simply with cold water. The *domus* itself transformed in the sixth century into a multi-family dwelling and the bath turned into a workshop (Perich Roca 2014). The baths of the Pati Llimona *domus* are smaller than those of Bisbe Caçador, but are very well preserved. They are also linked to a private house, which was also built on the *intervallum*. The bath with its hypocaust floor was in use until the sixth century, according to the abandonment layers (García-Entero 2005, 211).

The aqueduct

Sources, course and distribution

The aqueduct of Barcelona is partially still standing in the northern part of the old Roman city, exactly where they cross the walls next to the north gate, at the Plaza Nueva. It was thought to have disappeared long ago, and 19th and 20th century scholars argued about the location of the aqueduct, which was known from written medieval sources and from a Roman inscription:

> *L(ucius) Min[ucius L(uci) f(ilius) Galeria Na]talis (...) et L(ucius) Minicius L(uci) f(ilius) [Natalis Quadro]nius Verus f(ilius) (...) balineum c[um port]icibus solo suo et/ du[ctus aquae] fecerunt.*
>
> 'Lucius Minucius Natalis, son of Lucius, of the Galerian tribe (...) and [his] son Lucius Minicius Natalis Quadronius Verus, son of Lucius (...) built a bath with porticoes and an aqueduct on his own land.' (abridged, *CIL* II 4509)

Then, during the construction of a new public square at Barcelona (the Plaza Nueva) in the 1950s, the remains of the arches for two conduits appeared again inside the Casa del Arcediano (a Renaissance palace built adjacent to the town gate), projecting outside the walls (figure 35). It was evident then that the aqueduct had been included in the Roman and medieval fortifications (Fernández Casado 2008 [1972], 219, Mayer Olivé and Rodà de Llanza 1977, 271). In this site, two different water conduits are now standing, built in parallel and following the same construction technique.

They are built in *opus vittatum* with regular ashlar blocks quarried from the nearby hill of Montjüic. The pillars, which are 5 x 5 Roman feet (1.55 x 1.55m), are built on top of solid bases which rest on the soil. The pillars culminate in imposts which are 1.5Rf thick, on top of which rest the arches. The arches are round or voussoir arches, 5Rf in radius at the intrados (3.10m total width), and the voussoirs themselves are 1.75Rf high. The *specus* is built on a box lying just above the horizontal formed by the extrados, 5Rf wide and 3.5Rf tall, 7 metres above the ground. The western aqueduct, known as the Collserola aqueduct, has the biggest *specus*, is box-shaped, measures 0.6 x 0.6m and is lined in *opus signinum*. The eastern aqueduct, known as the Moncada aqueduct, is built with slightly smaller stones. Its *specus* is U-shaped, measuring 0.6 x 0.7m, and is also covered in *opus signinum*, but its level is 0.15m higher than that of the Collserola aqueduct (Miró Alaix and Orengo 2010).

Figure 35: Reconstructed remains of the aqueducts at the Plaza Nueva.

The existence of two different conduits entering the town has generally been taken as proof for the existence of two different aqueducts, but according to recent research both conduits are part of a single aqueduct which branched at some point between the source and the city (Orengo and Miró Alaix 2011).

The course of the main branch (Miró Alaix and Orengo 2010, Sánchez López and Martínez Jiménez 2016, 149-54) can be traced fairly accurately beyond the Plaza Nueva thanks both to archaeological data and written references, as is repeatedly mentioned in medieval documents (*archos antiquos*, *arcos*, *aquaria*, etc.) in the surroundings of Barcelona (Mayer Olivé and Rodà de Llanza 1977). The aqueduct turns north from the Plaza Nueva and is found again at the Plaza de Durán y Bas: when a house was demolished there in 2010, and once the debris was removed, a section of the aqueduct was uncovered inside the wall of the adjacent house. This section perfectly preserves four more arches (five pillars) and its *specus*. It is also built in the same 5Rf-moduled *opus vittatum*. Because it is still imbedded in the house's wall, and holding its weight, it is not possible to clear the infilled masonry in order to render it free standing. The next known section is located at the site of Calle Magdalenas 25, by an old nunnery. These remains were identified in 2005, and were part of the *specus* and its surrounding *opus vittatum*, which had been reused after Antiquity and were all covered in lime mortar. The next location of the aqueduct is known from three written sources in the current urban area of Barcelona: These indicate that the aqueduct was standing in 1763 inside a house at the area known as the *Archs de Jonqueres,* next to the Lesser Count's Palace in 1116, and by the De las Puellas (formerly, St. Peter's) monastery in 1044. Six kilometres away from the last attested location, at the site of Sant Andreu (Calle Palomares and Calle Coronel Monasterio), excavations have revealed more sections of the aqueduct. The remains were located in 2006 (Giner Iranzo 2007), and stretch over 90 metres, and were identified in several survey trenches and excavated in depth in three different sondages. They are built in *opus caementicium*, with a total width of 1.30m, the *specus* being about 0.6m wide. This is lined in *opus signinum*, on top of which sinter crusts between 2 and 14cm formed (this may indicate a very long period of flowing water, because the source of the aqueduct is naturally poor in calcium). Several *cippi* (markers) and *spiramina* have also been identified there. The source of the aqueduct is identified thanks to a final reference, dated to 987, which mentions the aqueduct next to a Roman road (*ipsa Aquaria Antiqua vel in via*), which from the context of the manuscript is identified as the road next to the Tapioles river and the Moncada mine. This mine was a source of drinking water in the Middle Ages, and can therefore be associated with the original Roman source for the aqueduct. Thus the aqueduct had an approximate length of 11.3km, and a drop of 18.12m.

The course (figure 36) given by the evidence for the Moncada aqueduct fits perfectly with the route of the main Roman road, running parallel to it for the most of its course, which would further confirm that it is a correct route. Furthermore, the aqueduct itself seems to be running parallel to a later medieval (11th c.) water conduit, the Rec Comtal, which fed water to the many feudal mills of the counts of Barcelona before arriving at the Count's Palace (built during the 11th c. on top of the old sixth-century episcopal complex) (Mayer Olivé and Rodà de Llanza 1977).

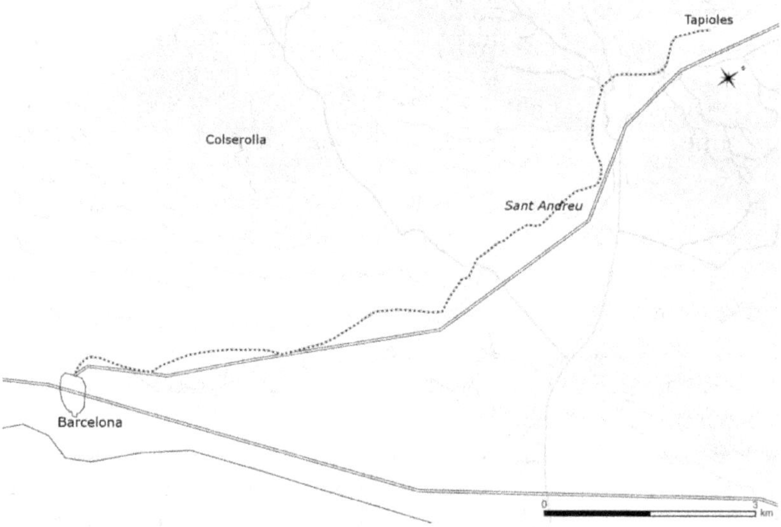

Figure 36: Reconstructed course of the aqueduct of Barcelona.

The other branch (which was thought to have been a different conduit) has been located archaeologically only in the area around the Plaza Nueva. From the standing remains, and heading north-west, several pillars were excavated at the Plaza itself, at the Colegio de Arquitectos, and then only identified later at Calle Arcos, whose name makes reference to the aqueduct itself (Sol Vallés 1977). All of these locations are only 200m away from the Plaza Nueva, so it is not easy to reconstruct the rest of the aqueduct's course.

Distribution and disposal of water

Inside the walls, the water seems to have been distributed from the *castellum*, which is thought to have been underground, by the forum, at the highest part of town. From there, water would have been supplied through pressurised pipes and open conduits to other parts of the town. One of these distribution conduits has been located at the Calle Palma de Sant Just, which at the time of its discovery (1904) was considered to have been the *cloaca maxima* or main drain (Fernández Casado 2008 [1972], Miró Alaix and Orengo 2010, Sol Vallés 1977). Another one has been identified leading towards the south late Roman fortified enclosure, misleadingly called *castellum*, where there was a set of baths (Ruiz Bueno 2018b, 59).

This conduit is a mortared-rubble, vaulted structure known for approximately 21 metres. It ran underground in Roman times, but recent studies suggest that this was not a drain, but a conduit of clean water. The arguments for this are its orientation along the *decumanus maximus* (from the forum terminal deposits towards the southern baths) and that it does not match other sewers of the city either in orientation or construction technique. Even if it is not lined with *opus signinum*, there are examples in Spain of aqueducts carrying clean water for baths (not necessarily drinking water) in clay-lined conduits, as in Pamplona (Unzu Urmeneta, et al. 2006), Valencia (see below) and León (Campomanes Alvaredo 2006).

As far as sewers are concerned, there are not that many known from Barcelona, but the most important for our study are those located at the episcopal complex. The old Roman sewers were completely reformed in the fourth century, but then seem to have been neglected and probably abandoned by the sixth century, but when the episcopal complex was developed, new drains were constructed. It is not clear if the drainage linked to the episcopal bath led to the main sewer or to a sewage-pit, but this drain was clearly built reusing the hydraulic systems of the pre-existing structures (Beltrán de Heredia and Carreras Monfort 2011).

Chronology and dating evidence

The aqueduct was built during the Flavian period, according to the material obtained from the foundation trench (Dragendorff forms 27b, 29) at the site of Sant Andreu, but what is really important is that there is plenty of evidence available from Barcelona to study the evolution of water supply during Late Antiquity (Miró Alaix and Orengo 2010). The aqueducts themselves are exceptional, as they seem to have been in use at least until the seventh century. The arches and conduits may have been standing for longer, as evidenced by the written documents mentioned above, with 34 mentions to the aqueducts between the years 987 and 1219, but this continuity does not necessarily mean that their use continued.

Focusing now on the late antique material, it is evident that the aqueducts were working in the fourth century: when the walls were expanded, the aqueducts that entered the city at Plaza Nueva were kept intact, and the tower built on the side of the gate was built aligned with the aqueducts, respecting their course and structure (Pallarés Salvador 1969).[4] Likewise, the water-consuming structures dated to the fourth century (the *cetaria*, the *fullonica*, and the new private baths built then), and the renovated sewer system indicate that running water was still available. The new baths constructed in the fifth century (Pati Llimona and Bisbe Caçador) further demonstrate this point. The material from the episcopal complex, expanded during the sixth century, also shows that the supply was maintained in the early Visigothic period. On the one hand, the baptistery might have needed running water, but this is not necessarily an indication of a continuing supply. The baptismal font is an octagon 4m across, and it has a drain to empty it of water, so it is conceivable that it was fed by piped water instead of being a bucket-filled structure.

What indicates a functioning supply in the late sixth and early seventh centuries is the new episcopal *balneum* (figure 37), which was built reusing the hydraulic infrastructure of the workshop area during the sixth century, as is evidenced by the presence of a large pool lined in *opus signinum*. The pool of the *frigidarium* was fed by a water pipe, as the imprint of the pipe has been preserved, which can be linked to the continuation of aqueduct water supply. Even

---

[4] This dating is based on similarities with other dated walls and the fact that they are built out of reused material, as no excavated material can date the expansion of the early imperial walls.

though only the *frigidarium* has survived, it is possible that it had hot rooms too, but these lie beyond the area of excavation. It has not been possible to date archaeologically the abandonment of these baths. The baths were discovered in the 1950s and partially dismantled, but in those excavations a latrine was also identified, which in the Roman period also worked with constantly running water (Beltrán de Heredia 2002c, 102, García-Entero 2005, 213). There is further evidence to suggest that the aqueduct was still working into the seventh century: at the Coronel Monasterio (Sant Andreu) excavation a fragment of an African-imitation Hispanic amphora (dated to the late-sixth/early-seventh centuries) was found mixed with other Roman material in the abandonment layers (context 120) of the *specus* of sector 100B, giving a *post quem* date for the context. There was another abandonment layer at this site, context 108, on top of the one with the late material, which contained largely Roman remains. From this very site, sinter deposits were obtained, which could provide further dating evidence through microscopic and isotopic analyses (Giner Iranzo 2007). The data from the second branch to the south '*castellum*' further reinforces this chronology, as it was clogged between the end of the sixth and the beginning of the seventh century.

There is no datable evidence for the use of the aqueducts beyond the seventh century. Afterwards, the available information relates to the actual abandonment and destruction of the aqueducts. The most important piece of evidence is a document dated to CE 1017, which reads as follows:

> *intus in civitate barchinona iusta ipsos archos priscos ... in praenominatis archis priscis unde* ***olim aqua consuerit decurrere.***
>
> (...) inside in the city of Barcelona next to those old arches (...) in the aforementioned old arches from which **water once used to flow** (Mayer Olivé and Rodà de Llanza 1977, quoting LAntiq. I 599, f. 222v)

This indicates that the aqueducts were out of use in the early eleventh century. The aqueducts had probably been pulled down during the course of the late ninth or the tenth century, which is when the Count's palace was built after the Frankish conquest, but it is impossible to tell if the aqueducts were carrying water then or if they were already out of use.

Figure 37: The episcopal balneum of Barcelona.

Discussion

There is certainly enough datable evidence to suggest that the aqueducts of Barcelona were in use, maintained, and preserved as late as the seventh century. The presence of this working water supply allowed the construction of a new set of baths in the sixth century, which must have been a source of great prestige for the bishop who built them. The repairs and the baths underline the position of the bishop of Barcelona as an urban leader in the Classical sense. This may also in turn reflect the political power hinted in the *De fisco Barcinonensi*, and displayed in the development of the episcopal complex.

It is not possible to tell if this aqueduct was properly working at full capacity at such a late date, or if it was only partially working. It is not possible to tell either when exactly the aqueduct ceased to function. The seventh-century amphora *terminus post quem* date for the abandonment layers at Coronel Monasterio, and the construction of the counts' palace in the late ninth, a *terminus ante quem*, leaving us with a two century gap in which the aqueduct presumably ceased to function.

The two main historical mentions of Barcelona in this period, the sieges by Wamba (*HWR* 11: *Prima enim ex rebellione omnium civitatum Barcinona in potestate principis religiosi adducitur* = 'Barcelona was the first out of all the cities of the rebellion which was put under the power of the pious prince') and by Louis the Pious (*VHI* 13), do not mention the aqueducts at all, or if they were damaged as a result of the sieges, but it is very doubtful that these events had any direct impact on the aqueduct.

BRAGA

Located at the northmost corner of Portugal, Braga was established as an Augustan foundation (*Bracara Augusta*) after the pacification of the Peninsula, and it served as a main centre of Roman power for this far corner of the *Tarraconensis*. As many other Augustan foundations, it was equipped with a large forum, baths, porticated streets, and a theatre.

In the late Empire it was promoted to provincial capital of *Gallaecia*, and this new status can be seen in the fourth-century monumental phase of the city. New colonnaded streets and further developments of the forum are a clear example of this (Fontes, Martins, et al. 2010), as are the new lavish aristocratic *domūs* dated to this period (Martins, Magalhães, et al. 2016). Probably at this same time is when the new set of walls was built, partially over previous Roman structures, suggesting perhaps that, as it happens in other cities across the West, the new, late enclosure was walling an area smaller than the original Augustan foundation (Lemos, et al. 1998, Ribeiro 2008). This regional importance made Braga an attractive prize for the Sueves, who took over *Gallaecia* in the early fifth century and established Braga as the capital of their kingdom (Díaz Martínez 2011). Despite the fact that Braga was the royal seat for over 150 years, all the evidence for royal Suevic monumentality seems to have been outside the city, at the acropolis of Falperra (López Quiroga 2004, 532-3) and the monastery of Dume (Fontes 1995), which sharply contrasts with Toledo and the Vega Baja suburban *praetorium*. The city was conquered by the Visigoths in the 580s and remained a provincial capital although, again, without any known or visible monumental input. This situation seems to have remained the same until the time of the Umayyad invasion, after which the archaeology has little to say. All throughout this period, the population seems to have shifted away from the forum, towards the corner of the walled city where the cathedral was located (Fontes, Martins, et al. 2010).

Most of our knowledge of Braga derives from the excavations of the sites known as the *insula* das Carvalheiras (a domestic context dated between the Flavian and the Suevic period) and the Alto da Cividade (a set of baths and a theatre), although the aqueduct has been identified and excavated at various different points (figure 38).

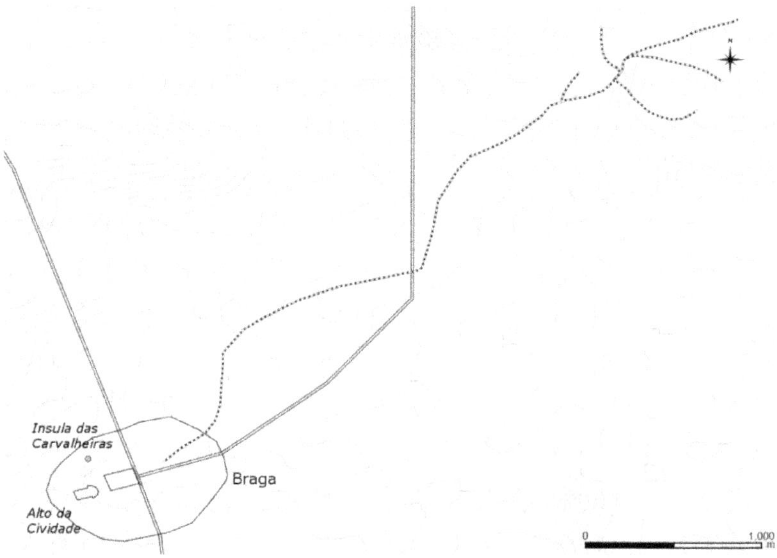

Figure 38: Reconstructed course of the aqueduct of Braga.

The aqueduct

The aqueduct has only been identified in urban excavations, whereas elsewhere outside the city the knowledge about the course is derived from name-place studies and other historical notes.

The exact location of the *caput* of is not known (Martins and Ribeiro 2012), but it is reckoned that the aqueduct obtained its water from the aquifer that springs towards the north-east of the city, because of its location, height and historical parallels. In fact, the 18th century archiepiscopal aqueduct obtains its water from various water mines, which may be Roman in origin. From there it might have run along the area known as the Sete Fontes ('seven fountains' or 'springs'), parallel to the Roman road and the Rua dos Chãos (the 'street of the pipes'), and in the area of the eleventh-century monastery of the Fontarcada (literally, 'arched fountain').

The *castellum* has been identified at the Alto da Cividade site, the highest point of the city (Martins 2005). From there a small branch went to feed the baths and another went to a secondary *castellum* by the forum. At the Alto da Cividade, a section of the aqueduct was excavated, built in mortared *opus vittatum*, with small blocks of granite rubble. The *substructio* seems to have been 1m wide, with a *specus* 0.6m tall and 0.45m wide, lined in *opus signinum*. On the opposite side of the city from the *castellum*, remains of hollow ashlar blocks, which if correctly identified, would relate to a suburban branch fed by a siphon (Morais 2011).

Further suburban remains, at the Law School, have been unearthed (Morais 2011). These, however, do not seem to belong to the main distribution network, and most likely relate to a suburban *villa* branch (Martins and Ribeiro 2012).

The dating evidence

The overall chronology of the aqueduct is difficult to determine, due to the nature of the evidence, as no datable materials were obtained from the excavated sections of the conduit. The Alto da Cividade excavations have unearthed a second-century pipe-fed fountain, so the aqueduct must be contemporary if not earlier (Morais 2011), which is consistent with the construction of the baths and the theatre (Martins, Mar, et al. 2013). For the abandonment we have further leads. These public baths were later modified and enlarged during the second and third centuries, confirming that the aqueduct continued into this period

Furthermore, both the Alto da Cividade bath building (figure 39) and the private *balneum* of the Insula das Carvalheiras were remodelled during the fourth century and in use into the fifth. The public complex was only dismantled and reconverted into a burial area after the fifth century (García-Entero 2005, 293, Martins 2005, 84-5). This fourth century functioning chronology for both baths is further supported by the hints towards other private bathing structures, such as those of the *domus* of Santiago (Martins and Fontes 2010, 118). Between the fourth and fifth century (again), new houses built around the theatre seem to respect the course of the aqueduct (Martins, Magalhães, et al. 2016, 40-50), while the theatre was itself already partly dismantled. All of this evidence confirms that the aqueduct was in use and in demand during the fourth century and probably into the fifth.

Overall, it is clear that the aqueduct was in use during the fourth century. It is equally clear that the local elites responsible for these baths and for the re-furbishing of the city and building of the walls was still very much present and active at the time of the Suevic invasion. Furthermore, the only positive evidence for abandonment and change of use for the Alto da Cividade baths are the post-fifth century burials. According to Hydatius (167), Theoderic II of the Visigoths sacked Braga in 456, but despite the many captives and sack, it was taken 'without bloodshed' (*incruenta...direptio*). As with other similar cases, the Visigothic conquest cannot be used as a terminal point for the aqueduct. All in all, it can be concluded, although without conclusive evidence, that the aqueduct of Braga was probably abandoned during the course of the fifth century, being impossible to refine the chronology further for the time being.

CONIMBRIGA

Located high on the Mondego valley, the site of *Conimbriga* is currently abandoned (figure 40), and it does not correspond to the current city of Coimbra, which is built on the location of ancient *Aeminiun*. Originally established as an Iron Age settlement (*oppidum*), which appears already under Roman control at the time of the campaigns of D. Junius Brutus of 138 BCE. It went through a main phase of monumentalisation in the Augustan period, and again in Flavian times. The site, which has been extensively excavated, includes a large forum-basilica complex, public baths, walls, an amphitheatre and many houses (Correia 2010).

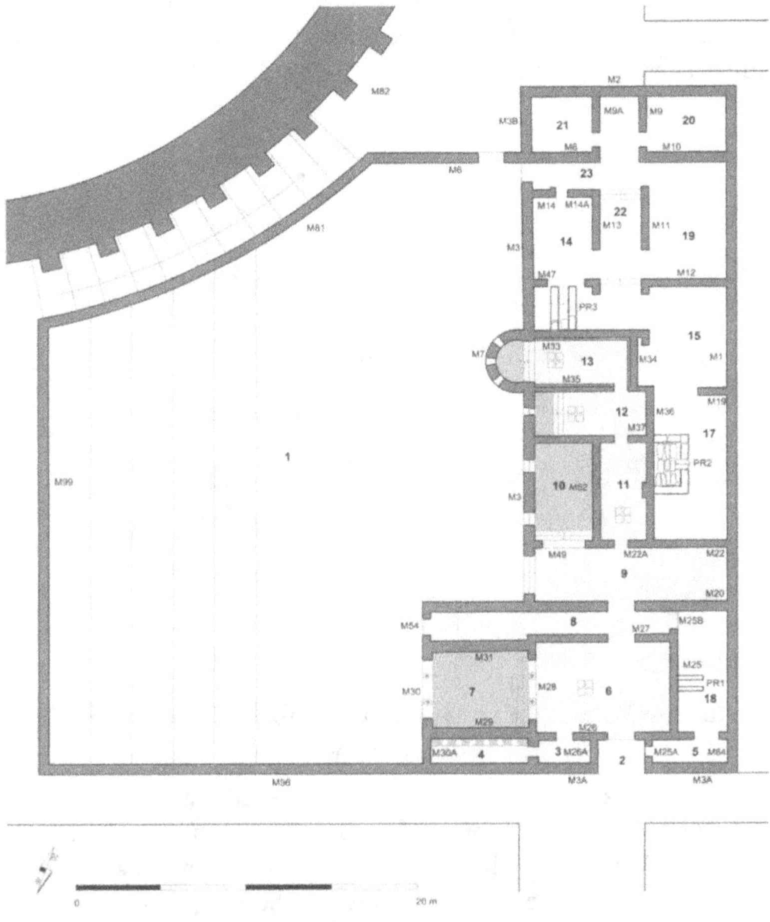

Figure 39: The baths of the Alto da Cividade in Braga (Martins 2005: fig. 16).

The late Roman phases of *Conimbriga* are characterised by the construction of a new fortification, which left a third of the old Augustan enclosure outside, including the amphitheatre, and cutting across various old *domūs* (De Man 2007a, 2011). And despite this, the city seems to have continued to flourish, if the proliferation of large houses with new mosaics and bath complexes is any indicator – the most famous of which is the *domus* of Cantaber (Correia and Reis 2000). The post-Roman period is an area which remains largely unknown, partly because of the text of Hydatius (225-7, 237), who mentions that the city was abandoned after the sack of 465, was used to date the violent destruction of areas of the forum and some *domūs* (Alarcão and Étienne 1977, 165).

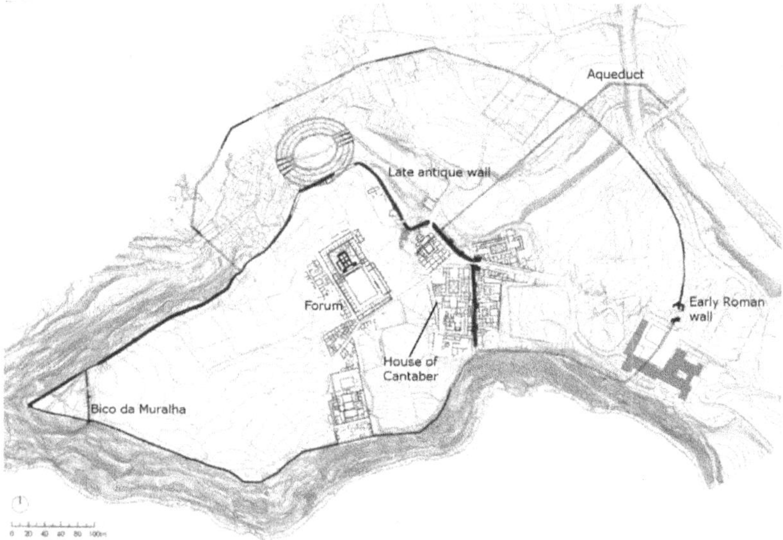

Figure 40: Plan of the remains at Conimbriga (based on De Man 2011: fig. 1).

More recent excavations have, however, revealed that habitation continued on site, although probably much more reduced in scale. The works by Adriaan De Man (2007a, 2011) have shown that the westernmost corner of the city was fortified in the Visigothic period, during the late sixth century, by means of the construction of a wall that enclosed a small space (the Bico da Muralha; figure 41). This late sixth-century citadel, has probably got to be related to the integration of that part of *Lusitania* in the Visigothic kingdom after the conquest of the Sueves (Alarcão and Étienne 1977, 165-8). A possible intramuros Christian burial site might have been identified, but the identification is dubious and its publication and excavation not up to modern standards (Correia, De Man and Reis 2011). Excavations in the area of the amphitheatre further unearthed remains of silos which have been dated to the Islamic period (Detry, Cardoso and Correia 2014), although no habitational structures were identified.

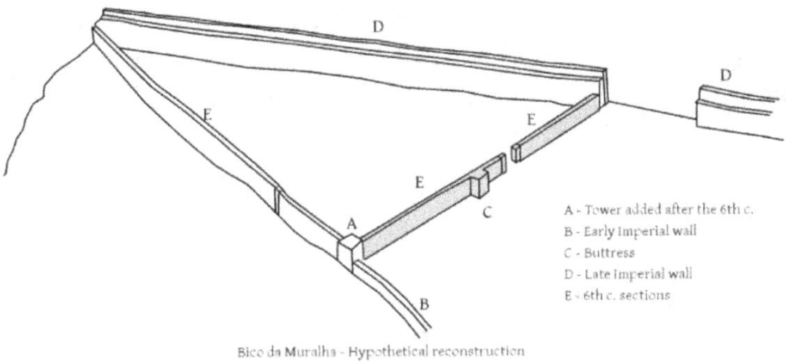

Figure 41: Bico da Muralha, the Visigothic fortified enclave in Conimbriga (De Man 2007b, fig. 5).

The aqueduct

The aqueduct of *Conimbriga* was surveyed and studied in the 1970s (figure 42), and its course was fully identified then (Alarcão and Étienne 1977, 52-7, Étienne and Alarcão 1974). Its course is known from its source up to its *castellum* inside the city. The source is a spring, which is dammed with a low (1.2m) wall and its overflow collected into a basin at the bottom of a tower. The *specus*, some 0.5m wide, is lined in *opus signinum*, runs inside a 1.5m-wide *opus caementicium* box, which is vaulted. From the *castellum* the aqueduct went to a system of twin *piscinae limariae* or settling tanks, which may have served to purify the water (Ohlig 2004). The distribution is done by means of lead pipes, which follow a geographical distribution. From there the aqueduct runs alternating underground sections and surface structures, so it is visible in 25 different locations. Just before *Conimbriga* the aqueduct had to go over arches for 108 metres, although sadly most of these are now gone. The *specus* was lined in *opus signinum* and the conduit had an average slope of 2.6m/km, the total length of the aqueduct being 3,443m.

The aqueduct was not excavated, only surveyed, so all the dating information available comes from indirect evidence, mostly the excavated baths of *Conimbriga*, although there is also the notice of a possible baptistery located beyond the east gate, which is the access point of the aqueduct to the city (Palol Solelles 1967, 169). The three public baths were excavated in the 1970s, and no abandonment layer could be identified or dated, although the Baths of the Aqueduct seem to have been used into the fifth century. Only one of the private *balnea*, that of the House of Cantaber (figure 43), has produced late antique material (Correia and Reis 2000).

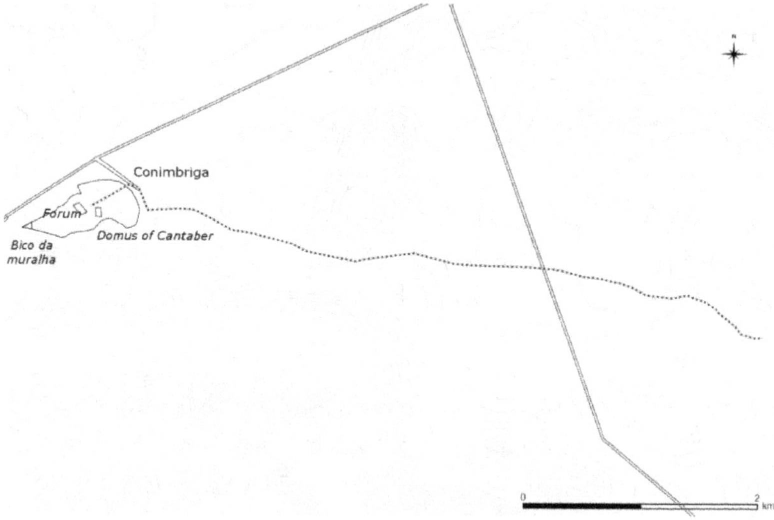

Figure 42: Map with the course of the aqueduct of Conimbriga.

The *balneum* of this *domus* seems to have been abandoned at some point early in the fifth century, because the hydraulic infrastructure of the baths appears to have broken and flooded the installation, which was then abandoned. This at least indicated that the aqueduct was working early in the fifth century. The house as a whole, together with its bath seems to have been destroyed in the Suevic attack, and it was not inhabited again afterwards (García-Entero 2005, 567).

If the city was largely abandoned after the Suevic attack, as it is assumed from Hydatius' text, then the aqueduct may have been out of use only because nobody was there to use it. The Visigothic occupation in the far west section, 120 years after the alleged abandonment, may have been supplied with water from the aqueduct if it still functioned, but probably (this is not certain, as it has not been excavated) it relied on cisterns.

Figure 43: View of the baths of the House of Cantaber (photograph by Filipe Portas, Wikimedia Commons).

CÓRDOBA

Córdoba was one of the key cities in Roman Spain.[5] It was the capital of the province of *Baetica*, the richest and most Romanised of all the provinces, with important contacts with the imperial family and a very strong and influential landed aristocracy linked to the production of olive oil. In Late Antiquity it lost some of its importance, as the provincial capital was moved to Seville, where the archbishopric was established.

Córdoba was, nonetheless, still a very important city in the early fourth century. The most important building of fourth-century Córdoba was the suburban palace-complex of Cercadilla, identified in 1991 when the train station was expanded.[6] This site includes not

[5] A recent summary en English has been published by Desiderio Vaquerizo and Juan Francisco Murillo (2016).

[6] Causing the complete destruction of the site (Marfil Ruíz 2000a).

only a large residence, which has been linked to the Tetrarch Maximian's (r. 290-305) family (Vaquerizo Gil and Murillo Redondo 2016), but was later expanded with a basilica, and may have come to serve as the early episcopal complex – although this is dubious and hotly contested. The complex included a set of baths too (the only known set of baths of late antique Córdoba), which may have been working up to the fifth or sixth century, which is the moment at which the palace as a whole seems to have declined (García-Entero 2005, 702). The entire complex was equipped from its beginnings with a new aqueduct.

During the fifth century, the city went through similar processes to other towns of *Hispania*: the streets and the forum were encroached upon by private buildings, the marble decorations of the theatre were dismantled and turned into lime in kilns, and it seems that it was sacked by the Rhine invaders during the early decades of the century (Hidalgo Prieto 2005, Ruiz Bueno 2015). The main sewers inside the walls seem to have collapsed and no longer been maintained properly, although again, the evidence for this is quite thin (Marfil Ruíz 2000b, 119).

Córdoba became during the mid-sixth-century a major source of problems for the Visigothic kings, because following the Byzantine invasion, the city refused to side with either Goths or Romans, and remained largely autonomous, especially after the rebellion against the Gothic king Agila (r. 549-554; *Hist. Goth.*, 45). This caused several sieges and attacks on the city, commemorated in coins (*Corduba[m] bis optinuit* - *CNV* #30). These appear to have resulted in great disruption in the urban layout: the destruction layers observed in the excavation of Cercadilla have been linked (without any real evidence) to Agila's attack on the city (Marfil Ruíz 2000b, 121-3). In this period, the episcopal complex is usually considered to have been moved inside the walled enclosure, as suggested by various inscriptions (*CIL* II^2 7.644, 7.693). However, its traditional location next to the Visigothic *palatium* (Marfil Ruíz 2000a), has recently been put in doubt with serious archaeological arguments (Arce Sainz 2015, Ruiz Bueno 2018b). This intra-mural palace was later turned into the Islamic palace, known in the early period as the *Ballat Ludhriq*, or the palace of Roderic (last of the Visigothic kings). The population of Córdoba decreased during this period, as the north part of the walled perimeter was abandoned, or at least became far less densely occupied, from then on clustering around the new cathedral-palace complex, as evidenced by the distribution of intramural burials (Hidalgo Prieto 2005, Ruiz Bueno 2017).

It is during the early Islamic period that Córdoba got completely rebuilt, when it became the capital of the new state, especially after 750. The status as capital meant many things, amongst them that the population grew, and newcomers were settled in newly built suburbs, like *Šaqunda* (Casal García 2008, Manzano Moreno 2006, 250-4). The new emirs were responsible for great construction works, which include the renovations of the Visigothic palace and the new Great Mosque with its forest of columns with superimposed double arches (Arjona Castro 2001, 69-81, Salcedo Hierro 2000, 16-8). Emirs Abd al-Rahman I (754-788) and his son Hisham (788-796) also repaired the city walls and the Roman bridge (Arjona Castro 2001, 67, 80). The palace and the mosque were further expanded during the ninth century, but the height of Córdoba's Islamic power was during the mid-tenth century, when Abd al-Rahman III proclaimed himself Caliph (AD 929). Córdoba was by then so packed and crowded that he even decided to build a new city 8km away from Córdoba, *Madinat al-Zahra* (Vallejo Triano 2010). The Umayyad rulers of Córdoba were very keen on reusing and maintaining Roman water conduits, and at least four major repairs/adaptations are known (Ventura Villanueva 2002).

The early Roman aqueducts

The proximity to Córdoba of mountains with many water springs prompted the construction of aqueducts by the Romans. Up to five different conduits can be identified of Roman or post-Roman date.

The aqueduct of Valdepuentes

The first aqueduct Córdoba had is still standing in several sections in the mountains west of the city, the most prominent remains being those located at Valdepuentes (literally, 'the valley of the bridges'), which has given its name to the whole of the aqueduct. Its course is very well known (figure 44) thanks to the surveys carried out by Ángel Ventura Villanueva in the 1980s and early 1990s, although these only showed the course of the aqueduct before it approached the city

(Ventura Villanueva 1993). The aqueduct has been linked to the Augustan conduit known from fountain inscriptions:

> *Aqua Augusta. L(ucius) Cornelius Serg(ia) aed(ilis) IIvir lacus siliceos, effigies aheneas de sua pecunia fecit.* (*CIL* II[2] 7.217-8)
>
> 'The aqueduct of Augustus. Lucius Cornelius, of the Sergia tribe, aedile, and duumvir built with his own money the stone fountains and the bronze statues'

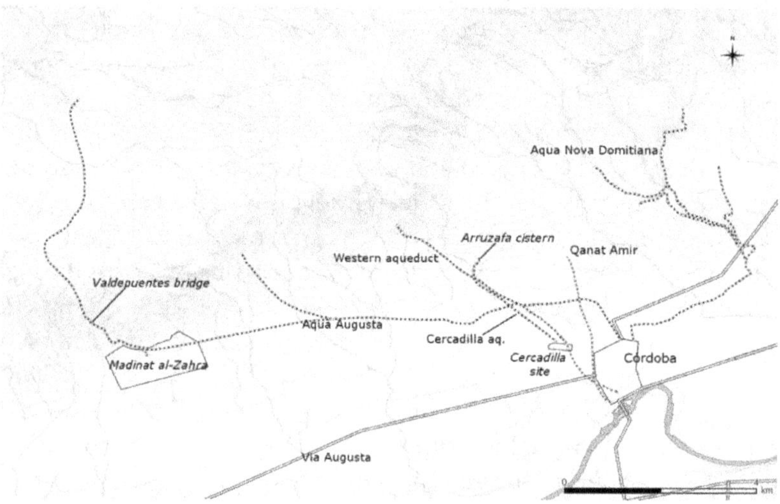

Figure 44: Map showing the Roman and late antique aqueducts of Córdoba (and, in red, the Islamic conduits).

In his first study, Ventura described in detail the aqueduct's course in five main sectors from its *caput* at the Bejarano river dam as far as *Madinat al-Zahra* and the Granja Agrícola. It runs mostly underground in its upper course (although some parts of it are built on solid walls), in a vaulted *opus caementicium* structure lined with *opus signinum*, and up to 0.64m across in the *specus*. Only when it had to go over a creek did the aqueduct go on arches. One of the most interesting features of this aqueduct is that, in order to bring the water down from the mountains to the plain in a short distance, 40 consecutive vertical drops (up to 25m apart) were constructed, allowing the aqueduct to descend 200m in just under 2km of its course (figure 45) (Ventura Villanueva 1993, 73-83).

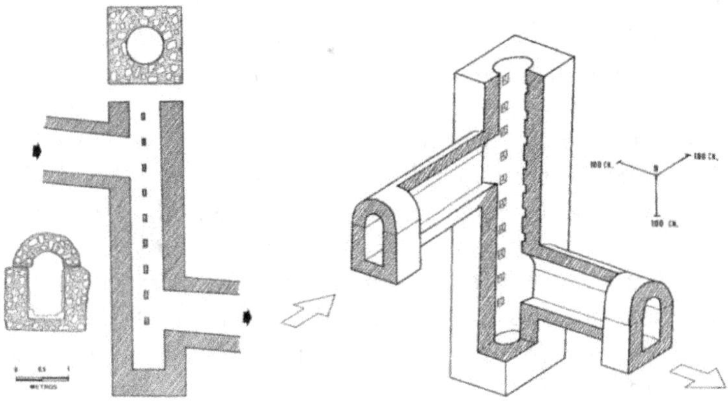

Figure 45: Aqueduct shafts designed to bring down water to a lower level without the need for an extremely steep-gradient conduit.

Since this work was published, many other remains of this aqueduct have been discovered archaeologically. In total up to 60 new sections, which ran on the surface at ground level until it approached the city, have been identified so far (Pizarro Berengena 2012, Ventura Villanueva and Pizarro Berengena 2010). The aqueduct seems to have entered the city over a long bridge of lofty arches, measuring up to 4.5m in height. The name 'Augusta' suggests that the construction of the aqueduct must have been Augustan, but not necessarily (Sánchez López and Martínez Jiménez 2016, 193). Later on (3rd c.?), a secondary aqueduct bringing water from the snow springs of Vallehermoso was added to this conduit to increase the volume of water.

The aqueduct itself appears to have collapsed during an earthquake in the third century: fallen remains have been found on top of late Roman floors. This is further reinforced by the ruptures of the conduit in sections 12 and 13, where the conduit was broken in different places and shifted from its original location.[7] Until this earthquake the aqueduct had been properly maintained, as there are no signs of sinter deposits inside these broken conduits. However, the

7 Numeration according to Ventura and Pizarro's 2010 study.

presence of calcareous concretions in some sectors of the upper course shows that the water was still flowing, even if it did not reach the city. The analysis of these deposits has demonstrated that the sinter had formed 39 annual-layers (figure 8), indicating that water was running through the collapsed aqueduct for 39 years before the *caput* blocked. Furthermore, the Cercadilla conduit, which was built to feed the tetrarchic complex, was built cutting through the Valdepuentes *specus*, rendering it completely useless thus dating the collapse and of the Valdepuentes conduit to some 39 years before that, in the 250's or 260's if we accept the chronology for Cercadilla (Pizarro Berengena 2012, 112-3, Ventura Villanueva and Pizarro Berengena 2010, 193-9).

The aqueduct was abandoned until the early medieval period when Abd al-Rahman III repaired and restored it in order to supply water to his new city of *Madinat al-Zahra*, as commemorated by this inscription now in at the Museum of Córdoba:

> '...] (...) The beginning of the works on this water conduit, from its very source, [was] in Shawwal of the year 328 (10th July until 7th August, 940). These works were carried out under the supervision of his (Abd al-Rahman's) client (*mawlà*), his vizier and City magistrate (*sahib al-Madina*) 'Abd [Allāh] ben Ba[d]r[...'[8]

The Umayyads reused the course of the aqueduct from its source as far as *Madinat al-Zahra*, including the shafts and wells mentioned above. New bridges were built when necessary, such as the bridge of Valdepuentes itself, and the new Islamic conduits were not lined in *opus signinum*, but in water-proof stucco (Ventura Villanueva 1996, 33-5). The bridge at Valdepuentes was actually built *ex novo*, after the old damaged bridge was pulled down. It is lavishly decorated on the side facing the valley (and it is not decorated at all on the side facing the mountain), and it is argued that it was the central decorative ele-

[8] Translated in Ventura Villanueva 2002 – This inscription has been linked to Madinat al-Zahra, although it has been suggested that it may have related to an aqueduct built to supply a private estate of the Caliph. This could also be the conduit mentioned by al-Makkari (III.4).

ment of the landscape, as all the mountain was a private hunting ground for the Caliph. The old Roman conduit was diverted into a new conduit outside the walls of *Madinat al-Zahra*, built in ashlar blocks and lined with red stucco (because the Roman conduit flowed too low to supply the Caliph's palace. The old conduit, however, was not fully abandoned, as it was reused at the same time as a sewer (Vallejo Triano 2010, 95-101).

The Arroyo Pedroche aqueducts

The second aqueduct that fed the city is known as the Arroyo Pedroche aqueduct because of the place where it was first identified. It seems, however, that its ancient name was Aqua Nova Domitiana, as shown by an inscription mentioning the name: *Aqua nova [[Domitiana]] Aug(usta)* (*CIL* II^2 7.219) the name of Domitian having been erased after his *damnatio memoriae*. This aqueduct has only been studied in detail by Ángel Ventura, in his 1996 work.

In it, he describes how the aqueduct did not have a *caput aquae*, but four. Water was collected from four different springs to the north-east of the city (figure 44), and was taken into Córdoba in four different conduits that merged into a single one. This unified aqueduct approached the city at ground level, and at some point was diverted by means of an inverted siphon, after which its course cannot be traced. The individual feeding aqueducts range between 3.4 and 0.9km in length, whereas the unified conduit is known for 2.9km, only 5m of which are actually on arches. The aqueduct is built in *opus caementicium*, but curiously, it is not lined with *opus signinum*, although the walls of the *specus* were lined with lime and clay (Pizarro Berengena 2012, 104-8).

Early imperial water consumption

These two aqueducts supplied water to the early imperial city (1st-3rd c. CE), and many water-consuming structures dated to this period have been identified. Unsurprisingly, most of these structures are generally not considered to have been working in the late Roman period (largely because many are old discoveries that cannot now be properly dated). These structures include seven *lacus* (public fountains), two baths, and up to fourteen private (domestic) fountains (Ventura Villanueva 1996, 94-125).

Two of these public fountains, located at Calle Ramírez de las Casas 13 and Calle Ambrosio Morales 4, were built in the early imperial period, dated by the inscriptions they bore (*CIL* II² 7, 217 and 7, 218) and were out of use in Late Antiquity (6th c. *t.a.q.*). The first one seems to have been dismantled then, according to the stratigraphy of the site, and the pilaster that held the spout was reused in a tenth-century nearby wall. The second *lacus* was also out of use by the Islamic period, as its inscription too was used in a wall dated to this period (Ventura Villanueva 1996, 95-8). A third fountain, located at Calle Maese Luis 20 (eastern suburb, fed by the Pedroche aqueduct), was excavated in 1993; it seems that the entire *vicus* surrounding the fountain was out of use by the fourth century (Ventura Villanueva 1996, 99).

Late antique water conduits

Besides the early imperial urban aqueducts, in Córdoba there were three later water conduits.

The Western aqueduct

The first of the late antique aqueducts is a suburban conduit, the Western aqueduct, discovered on the site of the modern Bus Station (hence its alternative name, the 'Bus Station aqueduct'). Excavations in the early 1990s revealed at this spot a 100m long section of a Roman aqueduct, built in *opus caementicium*, but surprisingly its inside lacks an *opus signinum* lining. The excavation also revealed the *castellum divisorium*, the distribution tank, where two different conduits met (figure 46) (Pizarro Berengena 2012, 102). This *castellum* is a small rectangular structure, lined in lead (in fact, re-using a lead sarcophagus), into which the water fell from the aqueduct. The *castellum* has two outlets that through two bronze pieces led the water into lead pipes, either for immediate distribution (for the suburban *vicus* and for the near-by circus), or to take the water elsewhere through long-distance pipes (Ventura Villanueva 1996, 118-20).[9] It

[9] Compare this distribution with that of the aqueduct of Metz (Lefevbre 1997).

seems, nonetheless, that this aqueduct supplied the suburban *vicus* west of Córdoba, as the walled city was already supplied by two other aqueducts.

Figure 46: Outlet of the Western aqueduct, identified at the bus station.

Its source may have been located at the springs of La Albaida (figure 44), some eight kilometres west of the city, and though the original Roman structure has not been identified, it is very probable that it is located below a late tenth-century structure (Moreno Rosa and Pizarro Berengena 2011). The route between the caput and the *castellum*, however, is completely unknown.

The construction of the aqueduct can be dated to the early imperial period by its construction technique; it is similar to other aqueducts of Flavian/Trajanic date, although the most recent theories suggest it is Neronian (Pizarro Berengena 2012, 99). It is also known that probably during the third century the aqueduct may have fed piped water to a *nymphaeum* built next to the Puerta de Gallegos, although it is not known for how long it was in use (Murillo Redondo, et al. 2002, 265). While there is no evidence to confirm that it continued working during Late Antiquity, we know it was working during the Islamic period.

The Islamic sources of the eighth and tenth centuries mention the existence of a fountain (*ᶜayn*) called *funt awrya*, which, according to Ángel Ventura, is not an Arabic name, but the preservation of a Latin one (*fons aurea*) referring to the bronze outlets of the terminal tank found at the bus station (Pizarro Berengena 2012, 104, Ventura Villanueva 1996, 2002). This led him, rather fancifully, to reconstruct the ancient name as the *aquaeductus fontis aureae*. More important than its name is the fact that a fountain was working in the north-western suburb in the early Islamic period (according to the sources), which may have been supplied by the aqueduct located at the bus station. This was then put out of use in 967 by Caliph Al-Hakam II, who redirected the Roman water source in order to feed the fountains and pools of the Great Mosque, which was until then supplied by a wheel:

> 'In this year of 356, water began to fill the tanks and the eastern and western ablution pools [of the Mosque]. It was pure water that came from a spring in the mountains of Córdoba, and looking for it the soil was excavated. It was brought [to the city] in a stone conduit, built solid and artistically, in which lead tubes were placed, so the water would not be polluted. Water first ran on Friday, ten days having elapsed in the month of Safar [25th January 967] (...)' (Ibn ᶜIdari, *Bayan*, II)

At the Bus Station, archaeologists have unearthed the tenth-century diversion tower which blocks the Roman conduit and the new Islamic aqueduct (figure 25). This is further confirmed by the presence of tenth-century glass vessels in the abandonment layers inside the Roman conduit and the lead-lined distribution tank (Carmona Berenguer, Moreno Almenara and González Viseda 2008). The text mentions that the water source was buried underground and that it was opened up and built again. This probably is the tenth century structure found at La Albaida. This Islamic conduit is still in use and supplying water to the fountains of the Cathedral (Ventura Villanueva and Pizarro Berengena 2010, 155-6).[10]

The Cercadilla aqueduct

Another suburban aqueduct was identified in a handful of excavations north-west of Córdoba (Ventura Villanueva and Pizarro Berengena 2010, 196-9), which has been linked to the fourth-century tetrarchic complex of Cercadilla, north-west of Córdoba.

Its source seems to have been at the snow springs of Arruzafa, where an *opus caementicium* conduit of Roman date has been identified, linked to a large water cistern (figure 47), which measures 77 x 5m and is lined with *opus signinum*. It has two construction phases, the earliest of which can be dated, according to the material from the foundation trench, to the fourth century (Murillo Redondo 2009, 463-71). This cistern is not the actual source of the conduit, which is located at some distance, but it may have served as a settling tank or a reservoir. Further remains were located at the Huerta de Santa Isabel, where a section 450m long was identified breaking through the old Aqua Augusta, which had collapsed in an earthquake during the third century. These remains where characterised by a lack of *opus signinum* lining, a narrow *specus*, and its brick and tile covering, all of which point towards a masonry case for a pipe conduit, the venter of a siphon (Pizarro Berengena 2012, 114-8). These remains follow a straight line heading towards Cercadilla, and used to feed the *nymphaeum* and the baths located at the complex – especially considering that the Valdepuentes conduit was already out of use by the time the palace was built. In the environs of Cercadilla, the remains of an *arcuatio* with at least 28 pillars has been found, also heading towards the palatial complex, although nothing is known about it regarding where it came from, where it headed to, or its chronology (Pizarro Berengena 2012, 118-22).

---

10 Indeed, a small manhole close to the archaeological remains of the bus station still marks where the water flows now, as indicated not only by the sign which reads "aguas de la catedral" (Cathedral's waters) but also by the constant sound of flowing water.

Figure 47: Arruzafa cistern, caput of the Cercadilla aqueduct (Murillo Redondo 2009, fig. 2a).

At some point after the fourth century, the complex appears to have been turned into a religious centre, perhaps an early episcopal complex or a martyrial shrine probably dedicated to St. Acisclus (Hidalgo Prieto 2002, 359-9).[11] By the sixth century, the baths of the palace ceased to function, as the rooms were reused as burial spaces, and all marble decorations and lead pipes were stripped off. This dating is further confirmed by the presence of African red-slip wares in some of the tombs, in particular form Hayes 104a, dated between 530 and 580 (Hidalgo Prieto 1996, 81, Fuertes Santos, Carrasco Gómez and Hidalgo Prieto 2013). During the mid-sixth century the Cercadilla/Saint Acisclus complex was used during the Visigothic civil wars as a military camp (*Hist. Goth.*, 45), so it is conceivable that this was done because there was available water for troops and soldiers (either through the Cercadilla or the Western aqueducts). The *castellum* of the aqueduct, however, seems to have been in use into the late-eighth century, when the large cistern was abandoned and its input source

[11] This identification is, however, contested (Marfil Ruíz 2000a, 157).

tapped to feed a private estate Abd al-Rahman I (Murillo Redondo 2009).

These two different abandonment dates may not necessarily have to be in conflict, as it is conceivable that the baths were out of use but the fountains and *nymphaeum* were not. This is a very tenuous suggestion and is only backed by an event which the Islamic sources record as taking place during the siege of 711. Various sources (Arjona Castro 2001, 21-7) mention how some Visigothic soldiers managed to survive entrenched at the basilica of Saint Acisclus because they had access to a conduit (referred to as *qanat* and *saqiya*).[12] None of the sources that describe this event are contemporary, and were compiled much later, which undermines the credibility of this episode. Furthermore, Guadalupe Pizarro (2012) has this episode taking place at Cercadilla but related to the Western aqueduct which was certainly in use and also in the vicinity of the remains of the palace complex.

Overall, it appears that the Cercadilla conduit was in use until the sixth century, if not straight up to the late eighth century.

The Alcázar conduit

The last conduit that should be mentioned is the Alcázar conduit, of which very little is known, except that it was built in the emiral period.

This conduit has been identified at two different locations next to the western wall of Roman Córdoba, at the Puerta de Gallegos (figure 48) and at the Vial Norte excavation, both sections follow the same route north-south. It thus comes from the north, and it has been suggested that it went directly to the Alcázar, the citadel located on top of the old Visigothic *palatium*. This links this conduit with that known in the sources as the *Qanat amir* (the aqueduct of the

[12] The whole story includes a black slave being sent by the Umayyad troops to investigate why the Goths will not surrender, who then gets captured by the Goths who believe him to be a painted soldier, so they throw him into the fountain and 'scrub him with water and a hard brush until he bled'. Upon escaping the slave told about the aqueduct, which the Umayyad soldiers then proceeded to block.

emir), built during the eighth century to supply water to the old Visigothic palace, by then turned into the *alcázar* (Pizarro Berengena 2012, 137-40).

Figure 48: Remains of the *Qanat Amir* at the Puerta Gallegos site, in Córdoba (Murillo Redondo, et al. 1996: fig. 2).

Archaeologically, the chronology that can be given is post-Roman, because it is sitting on top of late Roman (3rd/4th c.?) mosaics, and pre-Caliphate because it lies below tenth-century levels (Ventura Villanueva 2002, 125). No other datable material has been obtained from the excavations. The conduit itself is built using new (not re-used) material, put together in mortared rubble, and its 45cm wide *specus* is lined in *opus signinum*.

This conduit was still working in the early tenth century, as its fountain of piped water was still functioning, according to this fragment of the *Chronicle of al-Nasir* 28, which discusses how the water overflow from his private supply was used for a public fountain:

> 'In the year 306 [14th June 918 – 2nd June 919], al-Nasir li-din Allah [future Caliph Abd al-Rahman III] gave order to (...) build a pool next to the Fountain of the Pipe, located at the entrance of the palace, next to its gate, known as the Transenne Gate. The work was done with much care, and three water mugs were add-

ed to it, for the comfort of those who sought it. It was finished within a year (...)'

Discussion

With five main water conduits (two them working in Late Antiquity and a new one built in the eighth century), the city of Córdoba is certainly one of the clearest examples of continuity of water supply systems, and the dating for these structures is consistent. However, whereas the dating for the Roman period is clear, the dates for the Visigothic period are not that certain. As has been said: 'for sixth-century Córdoba, we deal with a field full of hypotheses which are difficult to confirm'.[13]

It can be claimed that, overall, public water supply ceased to function in intra-mural Córdoba by the fourth and fifth century, as the collapsed fountains and sewers indicate, even if they were later reused in the Islamic period (Sánchez Velasco 2011, Ruiz Bueno 2018a). This can be linked to the collapse of the Aqua Augusta after the 250's earthquake, although it is not possible to date the end of the Aqua Nova. Water supply inside the walls must have thus relied on wells or cisterns, even for the new late antique intramural baths (Ruiz Bueno 2017), as seems to have been the case during the Islamic period (Ventura Villanueva 2002, 123). The only structures supplied with water inside the wall enclosure were the Visigothic (later Umayyad) palace and the Great Mosque.

It is only in the Western suburb that running water continued to function during Late Antiquity, and this was supplied into the Islamic period by both the Western aqueduct and the Cercadilla conduit, even if the Visigothic continuity of the latter is arguable. These are dates supported by the archaeology and the information given in the sources.

The Umayyads inherited the urban water supply system, which they repaired and re-adapted to suit their own agendas. The re-use, repair, and maintenance of these structures took place largely during

[13] '*Para la Córdoba del s. VI nos movemos en un campo lleno de hipótesis difíciles de confirmar*' (Pizarro Berengena 2012, 127).

the early Emirate (second half of the eighth century) and in the early Caliphate (mid-tenth century), which were the moments of greatest urban construction (the Great Mosque, the palace, *Madinat al-Zahra*, etc.). The attitude of the new Islamic governors towards public water supply, however, was the opposite to that of the original Roman builders, as the water system was intended for the private use of the Umayyads (reuse of Valdepuentes for *Madinat al-Zahra*, Alcázar conduit) or for public buildings closely linked to the rulers (like the fountains of the Mosque).

LISBON

The city of *Olisipo* was an Iron Age settlement, which allied with the Romans during the second century BCE wars against the Gallaeci. It was promoted during the Augustan period to a *municipium civium Romanorum* (*CIL* II 175), with the title of *Felicitas Iulia*, and it remained a main commercial harbour for *Lusitania* and Mérida throughout the Roman and late antique periods. This commercial prosperity and wealth can explain the various raids and sieges during the fifth century, as the Roman authorities, the Sueves and the Visigoths fought over its control. Conquered by the Umayyads in 714, it still remained a rich and important city, being sacked by the Asturians as early as 796 and, during the ninth century, by the Vikings. Even if the archaeology cannot provide us with a detailed view of Ancient Lisbon, the mentions in the sources, together with a series of significant key rescue excavations allow us to get a general idea of its evolution during Late Antiquity. Its aqueduct can be dated to the third century, but it seems to have continued into the Visigothic period.

Late antique Lisbon

The Roman city, of which little is known (Fernandes 2013), seems to have been structured in three areas: an upper city with an acropolis, a lower city surrounding the forum, facing the sea, and a harbour town between the walled enclosure on the hill and the natural harbour (figure 49). It is suggested that the forum was under the cathedral although there is no solid evidence for this. The city also had a theatre, which has been thoroughly excavated. Lisbon also had an amphitheatre and a circus outside the city walls, which are only partially known. The circus at least seems to have been abandoned and reused for dwellings in the late third or early fourth century (de Sepúlveda, et al. 2002). There was, furthermore, a large set of public baths, the *Thermae Cassiorum*, which were renovated in the fourth century as commemorated in an inscription by the governor, Numerius Albanus (*CIL* II.191). Most of the lower city near the harbour was devoted, however, to the production of *garum*.

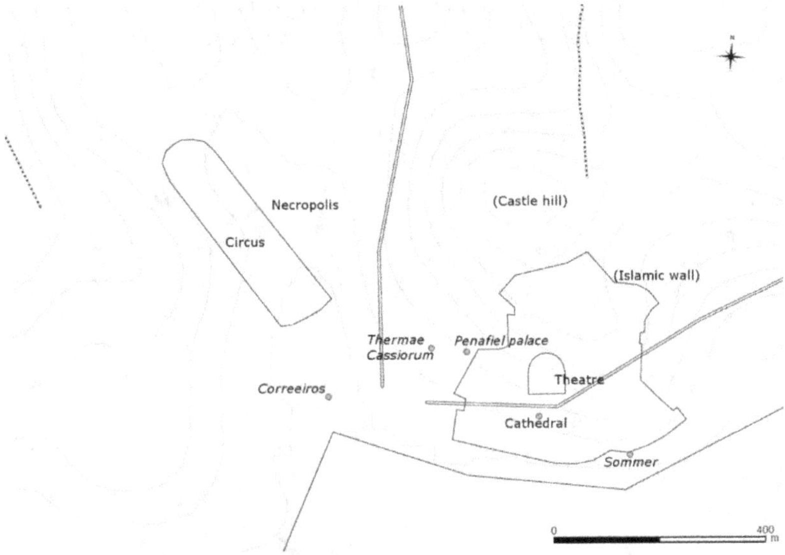

Figure 49: Main sites from Roman Lisbon.

During Late Antiquity the city shows similar patterns of transformation as other Hispanic cities. The wall enclosure, for instance, seems to have been reduced from its original Augustan layout: the V-shaped ditch excavated at the site of the palace of the counts of Penafiel suggest that the Islamic walls follow a reduced late enclosure (Silva and De Man 2013, Burgalhão 2009). These excavations have also provided with large amounts of pottery that underline Lisbon's commercial activity into the late sixth century connecting the Atlantic and Mediterranean routes (Fernández Fernández 2014), further confirmed by the presence of Byzantine coins. The *cetariae* and fish processing factories, which had been a very important part of the economy of the city in the Roman period (Filipe and Fabião 2006/2007), seem to have been out of use during the fifth century, as by the sixth some of them had been transformed into a necropolis (Casimiro and Silva 2013).[14] One of these complexes, at the site of Rua dos Correeiros, has a late Roman bathing complex, fed by a pipe, which shows continuity into the fourth century.

The Christianisation of the urban fabric again mirrors similar developments in *Hispania*. The presence of funerary fragments with Christian epigraphy in the sixth-century ditch of the Penafiel palace site points towards a possible Christian suburban funerary basilica. Similarly, it is guessed that the late antique cathedral was located under the eleventh-century mosque (Burgalhão 2009).

Figure 50: Roman dam, possible caput of the Lisbon aqueduct (Mascarenhas, Bilous and Neves 2012: fig. 2).

From a historical perspective, Lisbon was one of the main objectives of the Sueves, who in two occasions took and sacked the city, in 457 and 467 (both by treachery; Hydat., 181 [188], 240 [246]). Its wealth

[14] Although other *cetariae* in the immediate surroundings (but not in the actual suburb) of Lisbon, appear to have continued into the mid-sixth century (Bernal Casasola 2018, 112).

was also reflected in its political importance during the Visigothic period, when it was an active mint and the seat of a bishop.

The aqueduct and its chronology

The Roman water supply to Lisbon is very unevenly known. While the location of the *caput* is clear, and the upper sections of the course are visible still, there are no remains identifiable within the city. It was first noted on the 1960s (figure 50), when a large dam was identified 10 km away from the city (Sánchez López and Martínez Jiménez 2016, 238-40).

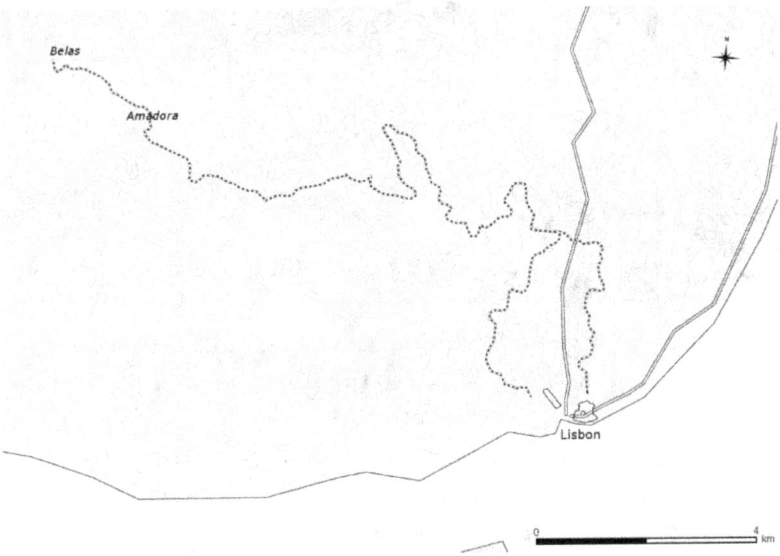

Figure 51: Reconstructed course of the Lisbon aqueduct.

The dam of Belas is some 15m long, 8m high and 7m thick, and it is built in *opus incertum* with various buttresses against it, across a river gorge. Calculations suggest that it could contain up to 125,000m³ of water. From here the conduit went on an over ground *specus*, lined in *opus signinum* and 0.4m across, following the contour line and gently heading towards Lisbon. The aqueduct has been identified at various points in the area of Amadora between the dam and the villa of Quinta da Bolacha (Mascarenhas, Bilous and Neves 2012, Fortes 2009). From there on the aqueduct's course is reconstructed with GIS analysis and following the known course of the Early Modern conduit. It has been proposed that the aqueduct had two branches, one leading towards the upper town (where the medieval castle of Saint George is located) while the other forked out and went towards the opposite hillock (figure 51).

Figure 52: Large late antique water cister, built with reused material, and potentially linked to the abandonment of the near-by fountain.

The excavations and surveys of the few known remains have not provided with any clear dating evidence. Only a suggestion that it may originally have been third-century in date has been put forward. For the abandonment, however, we have further hints, although none of them independently conclusive (Sánchez López and Martínez Jiménez 2016, 239). First we know that the large public baths, the *Thermae Cassiorum*, were still functioning and extensively repaired in the fourth century. Similarly, we have seen that the fish processing factories only go out of use during the fifth century (Lagóstena Barrios 2001, 44-6), although it is also at this time when a new public fountain was built at the site of Sommer (figure 16), which appears to have been abandoned during the sixth century (Mascarenhas, Bilous and Neves 2012, Silva and De Man 2013). In the same site, there was a large cistern (figure 52), built reusing architectural material, which co-existed alongside the aqueduct-fed fountain

(Ribeiro, et al. 2017). These two water structures were set up after the reorganisation of the area after the construction of the late antique wall. This overall suggests an aqueduct in use into the fifth century, and probably into the sixth (especially because of the Sommer fountain). Anyways, besides this late chronology, it is perhaps more important that until the (last?) moments the local council and the elites were engaged in preserving public access to piped water.

LUGO

The city of Lugo was an Augustan foundation (*Lucus Augusti*), which controlled a mountainous territory rich in minerals, one of the reasons why it played an important role in the late Roman period. Its strategic importance during the fourth century was further highlighted with the construction of an impressive set of new walls (figure 53), with 84 towers, enclosing 28ha of the Roman city (Fernández Colmenero and Rodríguez Cao 2012, Rodríguez Colmenero and Rodà de Llanza 2007). During the fourth century the city also reorganised and enlarged its sewer system (González Fernández and Carreño Gascón 2007, 259-60), and was an active pottery production site (Rodríguez Colmenero 2011, 87).

Figure 53: Roman walls of Lugo (photograph by Isaac Sastre).

Not much else is known about Lugo during Late Antiquity. Its strategic importance suggests that it was still an important city during the fifth century. The (manned?) walls appear to have kept the Sueves at bay, until they took the city by subterfuge in 460 – only to be sacked by the Visigoths later that year (Hydat., 199, 201). During the sixth century, and after the ecclesiastical reorganisation of the kingdom by Martin of Braga, Lugo was promoted to archbishopric, and we also know, from the sources, that at this period two large new churches were built, further highlighting its political relevance (Fernández Ochoa, Morillo Cerdán and López Quiroga 2005, 99, López Quiroga 2004, 47).

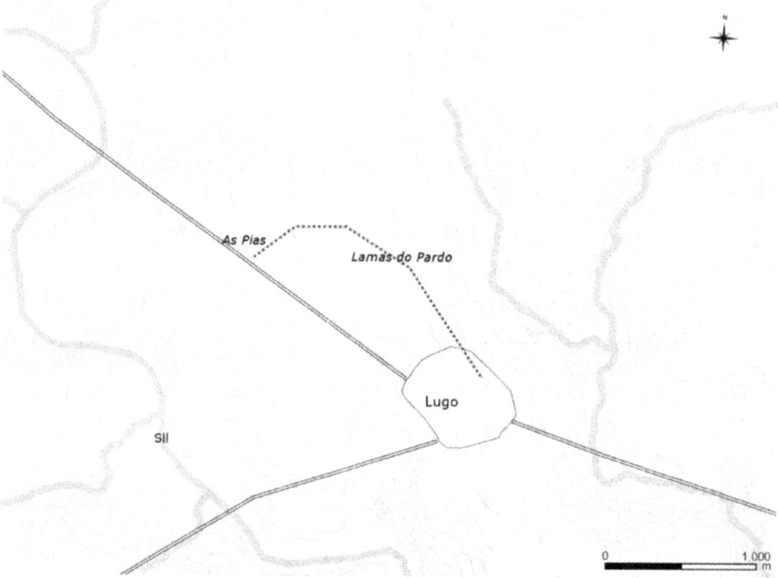

Figure 54: Map with the course of the aqueduct of Lugo.

The aqueduct of Lugo

The aqueduct of Lugo has been only very recently identified with certainty and excavated. There were seventeenth-century notices of a Roman water conduit, and in 1975 only two sections were known: the underground *specūs* of Rua das Norenas and Plaza de Santa Maria (figure 54) (Abel Vilela and Arias Vilas 1975). It was not until the late 1990s that new sections were identified, and the most recent discoveries date to March 2011,[15] which confirmed the course proposed in the first publication of the aqueduct, in 2003 (Álvarez Asorey, Carreño Cascón and González Fernández 2003, 20).

Considering this information, it is possible to claim that the aqueduct had its *caput* at the spring of O Castiñeiro, at the site of As Pias, 2.1km away from the distribution tank. The aqueduct is identified at some distance from the spring, at the site known as Lamas de Prado, where 8.7m of the aqueduct were identified in 2001, wholly built in *opus caementicium* and without any remaining hint of *opus signinum* lining (figure 55). In fact, the base of the conduit was tiled with bricks, which has lead archaeologist to argue for a piped conduit. Continuing downhill, two different *opus caementicium* walls have been recently linked to the aqueduct, although it is impossible to confirm this, as these sections (at Rua Juana la Loca and Rua Mazaria, where 200m are visible) are covered in thick vegetation. A small sondage at Camiño Real, not far away from the Rua Mazaria revealed an underground channel with the same orientation as the walls mentioned above; and not far away, at Plaza da Milagrosa, a 26m *opus caementicium* structure revealed the existence of the foundations for several pillars which would have supported the *arcuatio* of the aqueduct. The *specus* then ran underground, as identified in the 2011 excavations in front of the palace of San Marcos. The aqueduct would have finished at the distribution tank identified next to the old forum, at Plaza Santo Domingo, where a rectangular pool lined with *opus signinum* and linked to several lead pipes has been

[15] *Xornal de Galicia* (xornal.com) "Aparecen restos de un acueducto romano en unas obras en Lugo" and "Unas obras frente a la Diputación descubren un acueducto en Lugo", 8th March 2011; *El País* Galicia edition (elpais.es) "Los restos del acueducto de Lugo se abrirán al público en verano", 12th April 2011; *ABC* (abc.es) "Lugo mostrará en San Marcos el acueducto que traía agua a la ciudad romana", 6th June 2011. Compiled and briefly mentioned in Rodríguez Colmenero 2011.

excavated (Álvarez Asorey, Carreño Cascón and González Fernández 2003, 23-40).

Figure 55: Archaeological remains of the aqueduct (Rodríguez Colmenero 2011: fig. 36).

None of the excavated sections has provided any dating material that could date the abandonment of the aqueduct, so again, only through indirect information is it possible to roughly date the abandonment. It is important to avoid the Suevic conquest as an easy solution for this problem, not only because this is not accurate, but also because modern scholars consider that Lugo was destroyed (Abel Vilela and Arias Vilas 1975, García-Entero 2005), when Hydatius does not say so.[16]

The chronology

The baths of the city of Lugo provide very thin evidence, because two of them are known to have been built ex novo in the fourth century (as the baths of Plaza de Santa Maria, dated because of a mosaic excavated in the 1960s) or remodelled in the same period (as the baths of Rua de Armanyá) (Arias Vilas and de Vega Rodríguez 1997, García-Entero 2005, 250-1). It was also in the fourth century that the sewers were re-built, as mentioned above, and when the pottery workshops were still in use. The Bridge baths, however, seem to have been in use during the fifth century, when the pagan altars to the nymphs were destroyed and thrown into the pool, according to the pottery found in it. These baths, however, were built using the natural thermal water that sprung from the ground, rather than the water brought by the aqueduct (Meijide Cameselle and Herves Raigoso 2000). There is nothing more that can be said, as expressed in a recent publication by Antonio Rodríguez (2011, 228) 'considering our current knowledge, if we were asked if we would be able to tell the differences between the Suevic-Visigothic urbanism of *Lucus Augusti* and the original Roman one we would have to answer "no way",

[16] *Per Suevos Luco habitantes, in diebus paschae, Romani aliquanti cum rectore suo honesto natu repentino securi de reverentia dierum occiduntur incursu* = 'In the days of Easter in Lugo, a good number of inhabitants together with their leader of noble birth, feeling safe [because of] the holiness of the days, were killed in an attack by the Suevi'.

because the textual and epigraphic sources are virtually none and the archaeological evidence is as yet unexplored'.[17]

The water supply was certainly working in the mid/late fourth century, and it is conceivable that it was still working in the fifth, when we know, as mentioned by Hydatius, that there was still a strong elite in the city that could possibly have taken care of it. Without any hard evidence, however, it is not possible to make any claims on the date of the abandonment, despite unfounded claims which propose that the aqueduct may well have been in use until the Islamic invasion of 711 (Rodríguez Colmenero 2011, 229).

Figure 56: The Roman theatre of Mérida.

MÉRIDA

Mérida (*Emerita Augusta*) was founded as a veteran colony by Augustus early in the Principate. It was built with a pre-designed street-

[17] 'En el estado actual de nuestros conocimientos. Si se nos demandase diferenciar el urbanismo suevo-visigodo de Lucus Augusti del romano originario ¿podríamos hacerlo? De ninguna manera, tendríamos que responder, por cuanto las fuentes textuales y epigráficas son casi nulas y las arqueológicas inexploradas todavía'.

grid, walls, two monumental forums, and a north-east *spectacula* complex which included a circus, an amphitheatre and a theatre (Figure 56). The city is located between two rivers, the Guadiana (*Anas*) and the Albarregas (*Barraeca*), with a fertile land and a rich hinterland.

The city became the capital of the province of *Lusitania*, and during the late Empire it was also the capital of the *Dioecesis Hispaniarum*. During the fourth century the city prospered (Osland 2011, Alba Calzado 2018), as new large houses were being built (albeit partially encroaching upon the minor back streets). Only the suburbs west and east seem to have been in recession at this point. However, things changed abruptly in the fifth century. Archaeologically, it is possible to identify the dismantling of the forum and the street porticoes, the abandonment of the *spectacula* and a change in the housing patterns, where many of the old luxurious *domus* were subdivided into multi-family dwellings (Alba Calzado and Mateos Cruz 2008).

This sudden transformation is normally related to the attacks the city suffered early in the fifth century, it was first besieged by the Sueves in 429 and later captured in 439. Then Theoderic II of the Visigoths besieged it in 456 unsuccessfully, and it was finally captured by the Goths a couple of years later, after the battle of the Urbicus river (Hydat., 80, 106, 111, 126, 161, 163-8, 175). The city thus came under the control of the Visigoths, and during the reign of Euric, in 483, Duke Salla and Archbishop Zeno were responsible for the restoration of the walls and the bridge (*ICERV* 363). At some point during the sixth century, a large public building (a civic palace?) was built next to the forum (Ruiz Bueno 2018b, 100-1).

According to the *Lives of the Fathers of Mérida*, there was an archiepiscopal palace and a cathedral in the old forum, two other intra-muros churches and a large cult complex in the suburb dedicated to the martyr Eulalia. This suburban basilica has been excavated recently, confirming the data provided by the sources. Powerful archbishops such as Massona were responsible for several new buildings, like a *xenodochium* or pilgrims' hospital. This splendour visible in the written text is not yet matched archaeologically, where it is evident that during the seventh century the city's Roman infrastructure was in clear decline, as evidenced by the increasing amount of arable plots and animal pens inside the walls. There is one seventh century inscription, dated to the reign of Chindaswinth (r. 642-653) which may relate to a new building (*ICERV* 366, mentioned above).

On the eve of the Umayyad conquest the city still had a strong local aristocracy, which fought against the invaders after the battle of Guadalete. Even if the city finally surrendered, the locals kept rebelling against the emirs, so to prevent further uprisings, a citadel (*alcazaba*) was built inside the walled enclosure early in the ninth century.

The water supply

Mérida, located several kilometres inland in the flat valley of the Guadiana river, suffers from recurrent droughts, unpredictable rains and very hot summers, which made water supply a serious issue even back in Roman times.[18] That is why the city had to rely on various sources of water for its inhabitants: wells and cisterns are very common in the early Roman period and in Late Antiquity, but in order to supply water *en masse*, four aqueducts were built in the early years of the colony.

Los Milagros or Proserpina aqueduct

Of the four aqueducts of Mérida, that known as de los Milagros ('of the miracles') is probably the most famous. Its present name was given in the sixteenth century because most of the arches had gone, but the pillars were still standing, which seemed to be a miracle.

The aqueduct itself is known for 6km up to the Proserpina Dam (a name only given in the nineteenth century). However, recent LIDAR surveys have discovered a semi-subterranean conduit 50km in length which ends near the Proserpina Dam, and which might have fed into this aqueduct (Feijoo Martínez and Gaspar Rodríguez 2019). From the dam, the conduit runs in a rock-cut channel. Very

[18] Not only because of its scarcity, but also because of the suddenness and stormy nature of floods, which were periodical and very harmful, as narrated for the Visigothic period in the *Lives of the Fathers of Mérida* (*VSPE* II.21): '... the Guadiana flooded and having broken its banks, spread its waters far and wide, laying in ruins many buildings in the little villages by its stream...'.

early on, however, it approaches the city mostly above ground, and sections are known to have existed (or are still standing) at the site of Canja, El Sapo and the Carretera de Montijo (sections which measure between 20 and 130m), so its course is very well known (figure 57). In these sites, the size of the aqueduct and the *specus* seems consistent: 1.40m total width and 0.56m *specus* width, lined in *opus signinum* (Álvarez Martínez 2007, 198). Its *specus* was thought to have been open, but in the year 2000 excavations it was demonstrated that it was covered with a barrel vault for most of its course (Feijoo Martínez 2005).

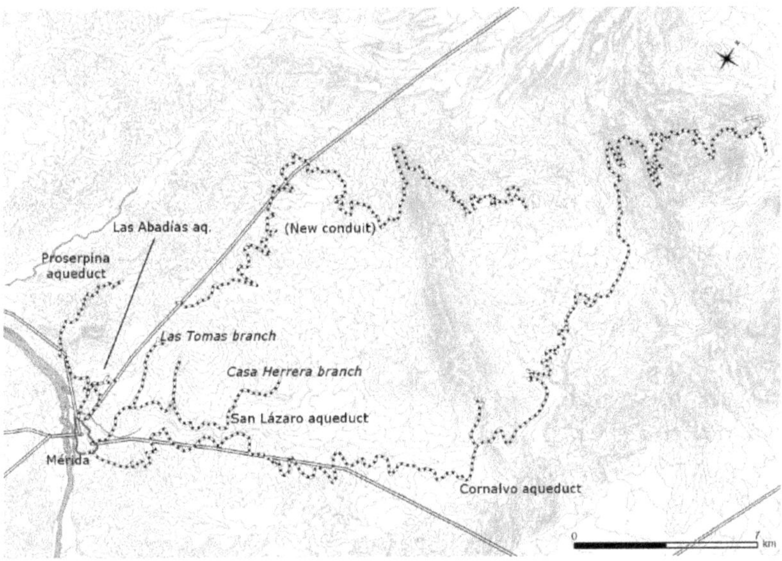

Figure 57: Map showing the aqueducts of Mérida, including the new sections identified in 2017.

The aqueduct enters Mérida by means of a 285 metre-long, 35 metre-high bridge (Fernández Casado 2008 [1972], 126-34). The pillars of this bridge are built with a core of *opus caementicium* and an ashlar facing. These are on average 7m apart, and their height varies between 35 and 15m. These pillars were originally 2.9 x 2.9m (10 x 10Rf) in section, but 2.1m (7Rf) reinforcement buttresses were added on the sides. The pillars are sustained by three rows of super-imposed arches, which alternate granite ashlar blocks with red brick, thus increasing the stability of the pillars (figures 5, 58). The bridge crosses over the Albarregas river, and the two pillars inside the river are reinforced with strong granite sterlings.

Figure 58: View of Los Milagros (Proserpina) aqueduct.

This section of the aqueduct begins at a settling tank and ends in a supposed distribution *castellum*, which nowadays is considered to also have been a *nymphaeum*, and just before crossing the walls it divided into two branches, one of which went to the north-west suburb, whereas the main branch supplied the low-lying western city (Mateos Cruz, et al. 2002, 74).

San Lázaro aqueduct

The San Lázaro aqueduct (also known as Las Tomas or Rabo de buey) is the second aqueduct of Mérida. Its course is the best known, because even if it was largely destroyed, its course was repaired in the sixteenth century (and still carries water) and because of the nineteenth century careful topographic study it received.

The aqueduct has three main feeding branches (comparable with the Arroyo Pedroche aqueduct of Córdoba) which obtain water from underground streams and from springs, known by the names of the springs they tap: Valhondo, Las Lomas, and Casa Herrera (figure 57). The three feeding conduits met outside Mérida and continue for 4km in an underground masonry channel, for which 99 *spiramina* have been located. When it comes to the surface, the conduit is carried on several arches built in *opus quadratum*, on top of which brick-and-rubble pillars were erected (figure 59), although sadly now most of the Roman arches have been substituted or destroyed by the sixteenth-century aqueduct (Álvarez Martínez 2007, 196, Fernández Casado 2008 [1972], 124-7, Gómez de Segura, et al. 2011, 134-5, Mateos Cruz, et al. 2002, 73).

Figure 59: Remains of the San Lázaro aqueduct.

This aqueduct entered Mérida crossing the Albarregas, close by the circus, the theatre and the amphitheatre, which may have been supplied by it, although the aqueduct must have also fed water to the eastern part of the city.

Cornalvo aqueduct

The Cornalvo aqueduct, or *Aqua Augusta* (according to an inscription located in recent excavations at the conduit), was the longest of the aqueducts of Mérida (Hiernard and Álvarez Martínez 1982).

Its source was considered to have been at the Cornalvo dam (Fernández Casado 2008 [1972], 117), but it is now known that the conduit begins beyond the dam, and it seems that an extra feeding branch was added to the aqueduct, diverting the spring of El Borbollón (Gómez de Segura, et al. 2011, 133). In 2017, LIDAR surveys revealed more sections of this aqueduct, which tap from springs in the mountains beyond the Cornalvo Dam. The total length of the aqueduct now is thought to be close to 75km (Feijoo Martínez and Gaspar Rodríguez 2019). The upper course appears to have been mostly underground or semi-subterranean (including vertical shafts like those of Valdepuentes in Córdoba), running in parallel to the Albarregas (figure 57). Along the valley, there are several points at which the aqueduct needs to go above ground: at Caño Quebrado ('broken conduit') one arch remains of what must have been a 30-arched structure and 30m of a wall have been located at Cerro Gordo.

Figure 60: Overground remains of the Cornalvo aqueduct.

The aqueduct entered the city from the east, next to the *spectacula* complex, and from there it went towards the southern part of the city, parallel to the wall, as evidenced by the excavations at the headquarters of the Guardia Civil, Villaemerita, Bodegones, and at Via Ensanche, where 85.7m of the conduit are still visible (figure 60). It seems as if the aqueduct fed the theatre and amphitheatre, although because it is the one which arrives at the highest point, this aqueduct may well have originally fed the whole city. Its name also would indicate that it might have been the first aqueduct to have been built (Álvarez Martínez 2007, 154-5, Hernández Carretero 2003, Mateos Cruz, et al. 2002, 72, Pérez Maestro 2005).

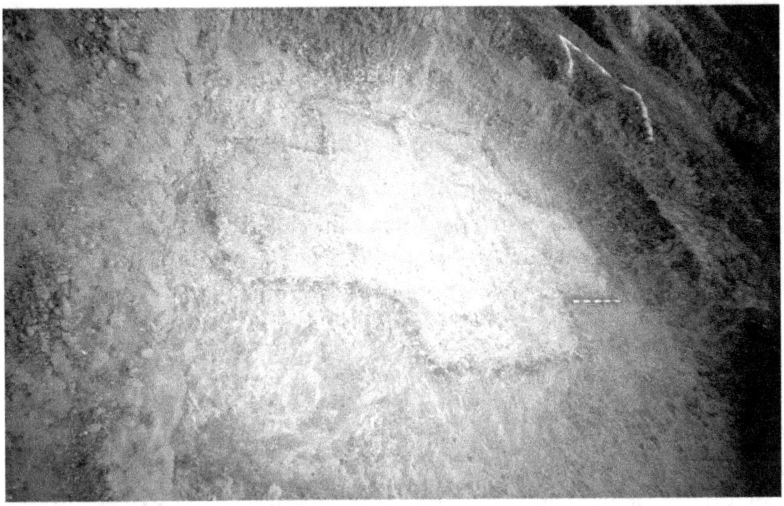

Figure 61: One of the pillars which has been identified as a foundation for the aqueduct of Las Abadías, at the site of the Miguel de Cervantes School (photograph courtesy of the Consorcio de Mérida).

Las Abadías aqueduct

The fourth and least-known aqueduct of Mérida (figures 57, 61), that of Las Abadías, was only identified in 2005, when the first remains were excavated (Sánchez López and Martínez Jiménez 2016, 244-5). The aqueduct is built following the same construction technique of the other aqueducts of Mérida, but its dimensions are different. The aqueduct has been identified in two different locations. The first one is at the site of Las Abadías, where 115.3m of the conduit have been excavated. Even if the conduit runs mostly underground, at some point it went over the surface, and four collapsed pillars were excavated at the site. The second location is Calle Medea, where a single pillar was later found.

Although it was originally thought that the aqueduct probably tapped water from Arroyo del Sapo, the newly discovered channel mentioned above might have fed into this aqueduct and not the Proserpina conduit (Feijoo Martínez and Gaspar Rodríguez 2019).

Dating evidence

Feijoo's thesis and the chronology of the dams

In 2005 and 2006, archaeologist Santiago Feijoo Martínez published in two articles his thesis which proposed that Roman aqueducts could not have been supplied by reservoir dams, and citing various authorities (Pliny, Vitruvius, Palladius, Frontinus) argued that water from reservoirs is not and was not drinkable. He draws a distinction between reservoir dams (that make a reservoir by blocking a river) and diversion dams, which simply took water from a flowing river, diverting it into the aqueduct conduit. The former were not suitable for human use, but the latter were, although their water may not have been the most ideal for drinking (as opposed to rain water and spring water).

Feijoo further argues that *opus signinum* was used in cisterns not only as a way of insulating the structure, but also because its pinkish-reddish colour countered the growth of algae and other microbes, which are greatly affected by that wavelength (their optimal wavelength being green-blue). However, recent studies on the quality of the water of these dams demonstrates that, despite slight pollution caused by animal faecal bacteria (which may not have existed in the Roman period, as that area was not used for extensive animal husbandry), the water is drinkable according to moderns standards, without the need for fluoridation, only needing filtering and decantation – purification techniques available to the Romans (Aranda Gutiérrez, Sánchez Carcaboso, et al. 2006).

This ground-breaking proposition directly affects our knowledge of the aqueducts of Mérida, in particular for two of them are generally believed to have been supplied by dammed reservoir water: those of Cornalvo and Proserpina (Los Milagros). Feijoo's arguments state that the dams could not have been the source of the aqueducts, because they would have carried non-drinkable water to a city located next to two rivers with easy access to abundant water for other purposes. This proposal is problematic, above all because it assumes that aqueducts were built to bring exclusively drinking water. Furthermore, his analysis of the text of Vitruvius is simplistic and acritical, as Vitruvius' text ought not to be taken as an accurate description of how things were meant to be. For instance, Feijoo argues that (2006, 146-7), according to Vitruvius, the *castella divisoria* gave maximum priority to public fountain supply and less priority to private individuals and workshop complexes, so aqueducts are necessarily supplying the same water to both drinking fountains and productive areas, which would make little sense if the water that ran in the conduit was not drinkable. However, Vitruvius' description of the *castellum divisorium* does not match those which are known nowadays, like those of Pompeii, Metz or Nîmes, which distributed water 'according to the geography of the areas served: pipes from the *castellum* carried water along main streets to designated neighbourhoods, and the same branch lines supplied both public basins and private homes' (Evan 1994, 8).[19] In fact, several aqueducts supplied the same water for workshop complexes and for dwelling areas, as the Aqua Traiana in Rome (Wilson 2000).

Then, Feijoo's study of the construction technique of both dams argues that the Proserpina Dam was built at some point in the early Middle Ages (7th-9th c.) and later expanded (figure 62), using other Visigothic and emiral buildings of Mérida as comparisons. He therefore proposes that the dams were built for agricultural or animal herding purposes once the aqueducts had ceased to function (2006, 157). Feijoo further criticises the scientific dating methods (radiocarbon date of a wooden plug at the base of the dam) which have been used to date the dam, disregarding them as unreliable because of the sampling methods and the way the data are interpreted partially without considering other datable materials from the same context. The archaeology of the building phases of the dam do clearly point to different construction phases.

[19] When it comes to aqueducts, there does not seem to be much trust in Vitruvius' descriptions (Fabre, Fiches and Paillet 1991a, 119-32, Lefevbre 1997, Brunn 2002b, 219).

Figure 62: Excavated remains of the Proserpina dam (photograph by Ángel Felicísimo, Wikimedia Commons).

On the light of recent discoveries using LIDAR surveying, it seems more likely that the dams were not part of the first Roman construction phase (Feijoo Martínez and Gaspar Rodríguez 2019). But still, the medieval chronology cannot be defended on the grounds of water quality alone. Dams could have been built in the Roman period at a later stage for multiple reasons, including problems with the original water source or an increasing urban demand (anything from new baths to a nymphaeum, the flooding of the amphitheatre arena, or the development of a new area industrial workshops). The construction of a dam to feed into the aqueduct would have ensured a constant and reliable water supply even in the summer months; a city like Mérida required water in quantity, and quality water could be obtained from cisterns and springs (both of which we know from the city itself). Urban aqueducts and low-quality water are not incompatible.

Further dating evidence

Independently from Feijoo's problem, in the past few years, there has been an increasing amount of available data from the excavated sections of the aqueducts in Mérida to propose a late antique abandonment. This contrasts with earlier publications, most of which argued for a fourth or fifth century abandonment (Osland 2011, 121, 128). The abandonment of the aqueducts has even been linked to the sieges and attacks on the city by Sueves and Visigoths during the first decades of the fifth century which, as it has been argued already, is too simplistic.

There is much debate about the phases and dates of the repairs of the Proserpina aqueduct, repairs which are clearly seen in some places where the construction technique changes. The interpretations of authors differ considerably: from Claudian, to Trajanic, Hadrianic, Severan and Constantinian, without a general consensus (Fernández Casado 2008 [1972], 137). Thermoluminiscence analyses of the bricks have revealed that those that were sampled can be dated to CE 288±85 [cal. 98%], which is not very helpful, but indicates a third-fourth century date for some of the repairs (Blasco, et al. 1993). After this, there is evidence for a late antique abandonment for this aqueduct: a bronze coin of Constantine was found inside the remains of a collapsed vault of the specus during the 2000 excavations outside Mérida, giving only a loose *terminus post quem*, considering the extended use of these coins through the following centuries, both for the collapse of this vault and the abandonment of the conduit

(Silva Cordero 2003). Besides this, there is a possible Visigothic repaired pillar at the Los Milagros aqueduct, identifiable due to its different construction technique, reusing material from the original fallen pillars but in a completely different way. This pillar is built out of the building material of other collapsed pillars but in a completely different technique (figure 9), perhaps indicating an attempt to restore the arcades which must have collapsed before (Alba Calzado and Mateos Cruz 2008, 267).

The Cornalvo aqueduct offers more precise dating evidence, as ten burials excavated inside the conduit at the site of Los Bodegones can be dated to the early Islamic period (Delgado Molina 2006). The necropolis was in use from the sixth century into the Islamic period, as indicated by an inscription in Greek, which has 'S[ub] EPA XIΔ', which should be read 'era 614', which equals to 576 CE (and not to 476 as the author claims). This, together with the construction of a large public bath complex behind it in the sixth century (figure 24) which was fed by a well rather than by the near-by conduit, may indicate a fifth-to-sixth century abandonment date.[20]

One of the feeding branches of the San Lázaro aqueduct, that of Casa Herrera, has been partially excavated, next to the site of the sixth-century rural church (figure 63).[21] The aqueduct was cut into the soil and covered by a masonry barrel vault. This section seems to have been working until the *specus* (lined in *opus signinum*, and measuring 40 x 45cm) was finally blocked with water-borne sediment (suggesting that the conduit slowly clogged up). Inside this sediment, fifth-to-sixth century material has been located, which matches the date of the construction of the basilica (c. 500), suggest-

[20] The date of this bath remains unpublished, even if it was excavated recently. I thank Miguel Alba Calzado, director of the Consorcio de Mérida for pointing this out to me during my fieldwork in Mérida. For earlier publications on the bath, see García-Entero 2005: 527-9.

[21] The excavation of this site was carried out in 2012 and 2013 thanks to the support of Lincoln College, the Craven Committee, the FECYT, and the Consocrcio de Mérida. The excavation was co-directed by Dr Isaac Sastre and myself, with the help of Valentín Mateos, Miguel Alba, Gilberto Sánchez, Pedro Dámaso Sánchez and Sara Rodríguez.

ing that the vault was opened in order to reuse the voissoirs in the foundations and get access to the still-running water. After the conduit finally stopped flowing, the spoliation trench was backfilled with soil. In this context a potsherd of the Mérida typology S-shaped rim (eighth-century in date) together with various other sixth and seventh century shards have been located (Cordero Ruiz and Sastre de Diego 2010, Gómez de Segura, et al. 2011, 139-41, Sastre de Diego and Martínez Jiménez 2012, 2013).

Figure 63: Excavated remains from the Casa Herrera branch of the San Lázaro aqueduct.

Besides the evidence from the aqueducts, there is plenty of information on water consuming structures in Mérida. The abandonment of the public fountains, for instance, is generally dated to the fifth century because supposedly it is then that the lead pipes were stolen (Alba Calzado 2007, 163). However, no dating material is presented from the robber trenches, and in reality only the assumption that this happened in the fifth century dates the recovery of the pipes. The abandonment of the only three structures which have been excavated that could be considered to be public fountains has been dated in this fashion, but no dating material has been provided (Mateos Cruz, et al. 2002, 77).

It is curious that the *Lives of the Fathers of Mérida* (IV.9.5) mention the city baptistery in the late sixth century. It was located separated from the cathedral in a 'little basilica' of St. John the Baptist (*basilicula*), which may have been a baptismal church, but still in the same episcopal complex, although we know nothing about where it obtained its water from.

Concerning baths, there are several private and public bathing complexes known from Mérida whose life can be dated into the fifth century and beyond. Of the private baths, the first example is the *balneum* of Calle General Aranda 15, which was excavated in the 1920s and was dated to the fourth and fifth centuries on the basis of the decorative motifs of its wall paintings and floor mosaics (García-Entero 2005, 558-9). The second example is the *balneum* of the Casa de los Mármoles, in the site of Morería, which was a private bath linked to a wealthy *domus*. It had two main phases, the last one spanning between the third and fifth centuries, when the site as a whole was destroyed by fire, being reoccupied in the late fifth century, according to the pottery, as a simple domestic structure (García-Entero 2005, 540-1). The last bath is the one excavated at Calle Nerja, which was located in the suburbs. It is a bath belonging to a late suburban *domus*, built in the sixth century and abandoned in the ninth. The bath itself is built using early imperial reused material. Its furnace has been dated through radiocarbon dating (from ash samples) to CE 607-694 [cal 98%], and in the foundation trench of the drain a potsherd with an S-shaped rim (dated to the eighth century) was uncovered. However, the bath seems to have been fed by a water wheel from the Albarregas river, rather than by piped water (García-Entero 2005, 560-1). Some other minor baths are known at the Train Station and in the Resti suburb, which during the Visigothic period ($6^{th}/7^{th}$

c.) were turned into houses according to the material retrieved from the fills of the vats. Of the various public baths known from Mérida (Osland 2011, 189-207), most appear to have gone out of use during the fifth century, partly dismantled and partly reoccupied. The bathing complex excavated at Calle John Lennon appears to have been rebuilt at some point after the fourth century, reusing the previously existing water supply, but there is not enough chronological evidence to refine the dating. Lastly, there is also the still-unpublished post-Roman bathhouse built by the theatre mentioned above, which was not fed by the aqueduct.

The sewers are also generally considered to have collapsed and been out of use by the fifth century, despite some minor and extraordinary examples of continuity into the Islamic period (Acero Pérez 2011, 2018).

A possible Umayyad reuse?

Al-Idrisi (1100-1165), in his work *The Description of Spain* mentions an aqueduct of Mérida. This was possibly the aqueduct of Los Milagros, on the grounds that this was the most impressive one and its location west of the city fits his description, although considering the orientation of Islamic cartography (where north is south), we may understand 'east' for 'west', which would rather point to the San Lázaro aqueduct. His description, written in 1154, may indicate that the aqueduct was in use:

> 'Amongst these constructions we will mention west of the city a great bridge with a great number of tall arches, whose top offers a wide passage. In the top fabric that overlies the arches there is a vaulted corridor which links the inner city with the far end of the bridge: it is possible to go through this without being seen. Inside this vaulted corridor there is a **pipe** which reaches the city. Men and beasts walk over [under?] these arches, which are of solid construction and notable quality'.[22]

[22] Al-Idrisi, *Description of Spain*, as cited in Fernández Casado 2008 (1972): 152 [my translation from his Spanish].

While the description of the aqueduct and its arched structure sounds perfectly plausible, the conduit is described as a secret passage into the city, and as enclosing a pipe. The mention of it being used as a passageway is already suspicious, as the conduit is not wide enough to walk through, suggesting this is either a literary topos or a confusion between the aqueduct and the bridge. The pipe must be, in any case, a later addition, because in Roman times the aqueduct carried the water in the *specus*, not by means of a pipe. This pipe may suggest that the aqueduct was put back into use at some point during the Umayyad period, just as the aqueducts of Valencia and Córdoba, although before the destruction and relocation of the rebellious population from Mérida to the new Umayyad foundation of Badajoz in 875. This has been taken as proof for an early medieval date of the Proserpina dam, which has two large pipe outlets at its base (Feijoo Martínez 2006, 151). With all this considered, nevertheless, the most recent approach to the sources (Elices Ocón 2017, 254-8) shows clearly that there is an overall confusion between bridge, aqueduct, legend, and topoi. The reference to the pipe is taken to be a description of the distant past, not of the twelfth century (when Mérida was largely abandoned) and no solid indicator for an Islamic phase.

Discussion

The dating evidence is very inconclusive. The overall picture seems to indicate that at some point during the early Visigothic period most water-consuming structures fell out of use, although this is a circular argument in which the aqueducts cease to function because there are no more water consuming structures, which are not working because the aqueducts are out of use. All this is based on the assumption that the fifth century attacks caused great damage to all four aqueducts.

However, considering the increasing number of private baths in the fourth century (Alba Calzado 2005, 128, 135), all of which suddenly seem to end in the fifth century, is this really a circular argument? Or is there some reasoning behind it? The city appears to have suffered greatly from the Suevic attack, when a whole section of the city seems to have been burned. I do not believe that the Sueves consciously pulled down the aqueducts; but it may be simply that after this the water-consuming infrastructure had been too damaged to be useful.

The Proserpina aqueduct was clearly abandoned by the Visigothic period (late antique coin in the abandonment layers, Visigothic pillar), and the Cornalvo aqueduct was most probably out of use during the sixth century (theatre baths, Islamic burials). A similar date can be proposed for the San Lázaro aqueduct, as the results of the excavation at Casa Herrera point out – even if the sixth century date would apply only to its suburban section, and cannot be confirmed for the city proper.

The fact that nothing is mentioned in the *Lives of the Fathers of Mérida* about water euergetism would clearly indicate that the aqueducts were out of use in the seventh century, although we still have to consider the possibility that there was an attempt to repair the Proserpina aqueduct. Perhaps it was easier to substitute the whole conduit for a piped supply, as it is inferred for the Umayyad period in Idrisi's text – if it really related a historical event.

RECCOPOLIS

Reccopolis was founded in the year 578 by King Liuvigild (r. 568-586), the foundation being best described in the words of John of Biclar:

> *Liuvigildus rex extintis undique tyrannis et pervasoribus Hispaniae superatis sortitus requiem propriam cum plebe resedit. Civitatem in Celtiberia ex nomine filii condidit, quae Reccopolis nuncupatur: quam miro opere et in moenibus et suburbanis adornans privilegia populo novae Urbis instituit.* (Bicl. 578.4)
>
> King Liuvigild, once the tyrants were destroyed on all sides and the invaders of Spain overcome, had his own peace to settle with the people. He founded a city in Celtiberia, which he named Reccopolis after his son [Reccared]. He endowed it with splendid buildings, both within the walls and in the suburbs, and he established privileges for the people of the new city.

The date is clearly given in the chronicle, and the reasons behind the foundation are clear: the city was founded as a celebration of the newly established peace in the kingdom, shortly before the rebellion of Hermengild. It is not clear, however, if this date relates to the inauguration of the city or the beginning of the works, as the city certainly took several years to build. It may have been also a way to celebrate the *decennalia* of his reign. Beyond the statement given by the chronicle, the foundation of the city was yet another step forward by Liuvigild to strengthen the Visigothic kingdom, by which the Visigothic king took over powers which *de iure* belonged to the emperor (like minting coins with his own name, fixing taxes, wearing purple robes, and founding cities). The city should not be seen as an empty act of propaganda, because it soon became a centre of glass and gold production, as well as a mint, so it is sensible to assume that it functioned as a major economic centre in the region (Castro Priego and Gómez de la Torre-Verdejo 2008, Olmo Enciso 2008, 43).

The city of *Reccopolis*

The excavations at *Reccopolis* have revealed two main areas (figure 19): an upper town, which consisted of a public enclosure, linked to the palace, and a lower town where houses and shops were located. The city wall with its towers have also been partially excavated, and encircle an area of 21ha (Gómez de la Torre-Verdejo 2008, Henning et al. 2019). The upper city, the so-called palace complex, consists of several buildings located at the north end of the city. Two long (120m), aisled, parallel buildings, built in mortared masonry, with tiled roofs and paved in *opus signinum* are the main structures, located at the north and south sides of the upper terrace. Both structures were double-storied and were separated by an open courtyard. The east end of the courtyard was closed by a small porticoed structure of similar characteristics as the two mentioned above and by a three-aisled cruciform basilica with an attached baptistery. The basilica is linked to the south main building by a monumental archway, which led to the main street going southwards, linking the palace courtyard with the rest of the city (Olmo Enciso 2008). Recent geomagnetic surveying has revealed a densely-built habitation, a street layout and more buildings potentially linked to the palace complex (Henning, et al. 2019).

The lower city, south of this arch, has only been partially excavated. The main street is flanked on both sides by workshops where glass furnaces and traces of gold-smithing have been identified, in what seems to have been the commercial centre of the city. It has been claimed that this pattern imitates that of Constantinople (palace complex – monumental archway – main street – city gate), and despite the huge distances that separate the two and the massive difference in scale, there may actually be some truth in where the inspiration came for the layout (Olmo Enciso 2008, 50-1). Behind these workshops several houses built around courtyards have been identified.

There are some water-related structures at *Reccopolis*, but none can be directly linked to the aqueduct. There is the baptistery mentioned above, built by the basilica, and there is also a big public cistern (figure 18), built on the main street and accessible through a paved porch, which might have been fed by the water channel identified at the palace area (Henning, et al. 2019, 742). Some drains are known as well. If *Reccopolis* was a royal foundation and a royal seat,[23] it would not be surprising to find a bath complex, as the Germanic kings seem to have been very fond of them, as testified by those of Ravenna (*Var.*, V.38), Pavia (*Hist. Lang.*, VI.20), Rome, or *Alianae* in Africa (Chalon, et al. 1985). Only 1.3ha of the site have actually been excavated, so there is always the possibility of still finding a set.

The chronology of the site is largely based on the archaeological record. Besides the mention by John of Biclar about the foundation (confirmed by the chronology of the coins deposited in the foundation trench of the basilica), nothing else is known of *Reccopolis* in the sources. There seem to be four main phases, including an early Umayyad one. The city was in decline from the mid-seventh century onwards, when its planned urbanism was abandoned and new structures were built encroaching onto the streets. In this moment, imported materials cease to be present on the site. By the end of the eighth century the city was largely substituted as a regional centre of power by the nearby Umayyad settlement (*jund*) of Zorita (which is where the modern village is). *Reccopolis* and Zorita struggled for control over the territory for some time, when the palace-complex seems to have been turned into a citadel, while the houses below were largely abandoned. This situation came to an end in the 850s, when the

[23] Arguments for and against this supposition are plenty in favour (Ripoll López and Velázquez Soriano 2008) and few against (Arce Martínez 2010).

site of *Reccopolis* was finally destroyed by fire, and abandoned until the Christian reoccupation of the area in the twelfth century (Olmo Enciso, Castro Priego, et al. 2008, Sanz Paratcha 2008).

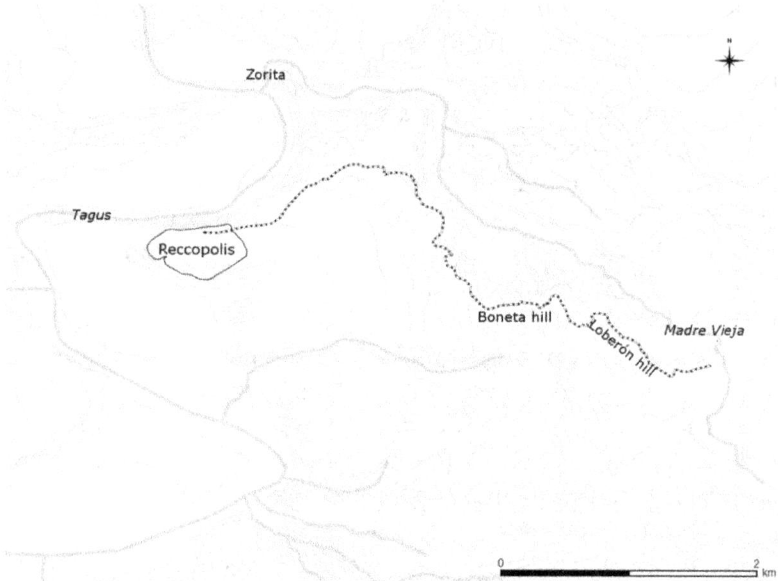

Figure 64: Map of the known and reconstructed course of the *Reccopolis* aqueduct.

The aqueduct

The aqueduct of *Reccopolis* has been known for several years, and mentions to it can be found in publications about the city, but no study of the remains had been published (Martínez Jiménez 2015, 2017). In 2010 and 2011, I carried out two field surveys in the territory of the site, and in the course of these two campaigns the aqueduct's course has been identified for over 2km in the hinterland (figure 64). These remains have been identified in the areas where the hill's slopes are the steepest, because in the other areas the whole of the mountain has been levelled by agricultural terraces after the Spanish Civil War (1936-1939), which destroyed all remains of the aqueduct. The remains which have been discovered are at some points badly damaged by pine trees planted in the early 2000s.

Figure 65: Remains of the aqueduct at the Boneta Hill.

The source of the aqueduct has not been located, despite attempts in 2010 to locate it around the many springs that can be found in the proximity of *Reccopolis*. Topographical observations in the area indicate that the source must be close to the Madre Vieja creek, a small river which carries water all year long and which is extensively used for agricultural purposes.

From the creek, a modern irrigation channel flows towards the north slopes of the Loberón hill, where four different sections of the aqueduct have been identified. After turning around the hill towards the south and then to the west, the aqueduct is next identified at the Boneta hill (figures 65, 66). The remains at the Boneta hill include a section excavated in the 1980s and the best-preserved standing remains, including a 2m tall wall, the remains of arch foundations and other ground-level *specūs*. From here, sporadic remains of *opus signinum* point towards the route of the conduit, following the contour lines, until a next section is identified at the Camino del Noguerón. This section is buried underground but was clipped by the bulldozer when the new road was excavated. A final wall is identified at La Paeriza, after which it has been impossible to identify any other remains.

The aqueduct is built in mortared rubble, with a *specus* lined in *opus signinum* of poor quality, using large chunks of tile and pottery. The conduit is 1.4m wide, and the *specus* some 45cm wide. The gradient of the conduit, as calculated at individual sections, varies between 22.7 and 21.3m/km. These gradients are much larger than those suggested by classical authors and very different to those found in other Spanish aqueducts, like those of Valencia (2.44m/km), Segovia (1.2m/km), or the Francolí aqueduct of Tarragona (4m/km) (Sánchez López and Martínez Jiménez 2016), although the overall gradient seems to have been much gentler, on average 1.45 m/km.

Chronology and dating evidence

At first, there were doubts over whether the aqueduct was in fact supplying *Reccopolis* or if it was Visigothic at all (Arce Martínez 2010), but the results of the survey confirm both its function and dating.

On the chronology of the aqueduct, the fragments of pottery retrieved from the *opus signinum* are of Visigothic date, according to their decoration and typology. Plus, much of the other material found in the environs of the structure also matches this chronology. The construction technique is also an indicator of Visigothic construction, in particular the use of mortared rubble and the coarse *opus signinum*, in which the fragments of pot have hardly been crushed and most of the material consists of tile. This, together with the

Figure 66: Further remains of the aqueduct.

fact that there is no other Visigothic site known in the environs, further links *Reccopolis* with the conduit, which allows us to propose a broad chronology ranging from 578 (foundation of *Reccopolis*) to c. 850 (abandonment of the site). It is not possible to give a more precise dating because it has not been possible to excavate the remains of the aqueduct and because the water consuming structures of the site (namely the baptistery and the great cistern) were either excavated in the 1940s without any right indicator of date being identified, or have not provided any dating material. However, the most likely date for such an unusual structure is the time of the royal foundation of the city. In fact, it could be the case that the aqueduct was built in order to ensure a constant supply of water for the construction site.

The sections of the aqueduct close to the source may have carried water for a longer period of time, and it seems that parts of the aqueduct were reused recently as irrigation channels. Locals mention stories told by the parents of the elders of the nearby villages (Albalate de Zorita and Almonacid de Zorita) in which the aqueduct (the 'wall of the Moors') still carried water, which may confirm that large sections of the aqueduct were only destroyed after the war, when intensive agriculture led to the construction of cultivation terraces.

SEGOVIA

Segovia is built on a granitic plateau in the Duero basin, 50ha in extension and situated 1000m a.s.l., between two rivers (which are at points 70m below the cliffs of the north-western end), making it only accessible from the south-east.

Our knowledge of pre-medieval Segovia is almost non-existent, even if paradoxically it has the most famous Roman structure of the Iberian Peninsula: its aqueduct. The historical centre is in its entirety covered by medieval buildings, and there have been very few excavations inside the upper town, although it is supposed that the original pre-Roman and the Roman settlements were both there, thus the need for an aqueduct. The archaeological information for late antique Segovia is, needless to say, nearly non-existent. Various villages and other rural sites and necropoleis are known (Tejerizo García 2017), but for the city itself the evidence is limited to one church. The church of Saint John of the Knights, currently a Romanesque building, appears to have been built on top of (at least) two late antique basilical structures, which can be linked, perhaps, to the promotion of the city to bishopric and mint during the sixth century (Sastre de Diego 2013).

The aqueduct

The course of the aqueduct is considered to be well-known, because the fifteenth-century reconstruction is still clearly visible from the source up to the beginning of the Roman *arcuatio*. However, excavations in the 1970's at one of the settling tanks have revealed that the 15th century and the Roman conduits did not, in fact, follow exactly the same route (Almagro Gorbea and Caballero Zoreda 1977). Despite this, this difference between courses at this specific spot may not have been typical, and it is safe to assume that the late medieval aqueduct followed roughly the original course, since this is the easiest one in practical terms.

Figure 67: The arcade of the aqueduct of Segovia.

Thus, the aqueduct begins at the Río Frío dam, and then continues for 9km at ground level through an open channel up to the forest of Valsaín, at which point the aqueduct goes underground, only to emerge close to the first settling tank, over 8km away from Valsaín. It then continues above ground on a solid wall until the second settling tank, from which the *arcuatio* begins (figure 67). The work on arches runs for 958m, in three different consecutive sections, with a total of 75 single and 44 double arches (Fernández Casado 2008 [1972], 56-87). From here, the Roman course is not known archaeologically, but if it were to follow the fifteenth-century route, it would continue underground up to the Alcázar, at the north-western end of the plateau (Zamora Canellada 2007).

The construction of the aqueduct can be securely dated to the late first century CE, thanks to several inscriptions. Some of them (*CIL* II 2739, 2746, 2751) are simply reused tombstones, which according to the type of lettering can be dated to the early first century CE. The most important of all, however, has not been preserved, which is the one situated on the aqueduct itself, just before it enters the city. It has been reconstructed several times based on the position of the clamps that once held the bronze letters, the most accepted reconstruction being that of late G. Alfőldy (1992). The inscription, according to this optimistic reconstruction, is dated to Trajan's second consulate, in 98 CE, and commemorates a restoration of the aqueduct.[24] This is further suggested by the presence of a coin of Trajan in the mortar of the *specus* (Almagro Gorbea and Caballero Zoreda 1977, Zamora Canellada 2007). This interpretation, however, is not widely accepted, and there are various other alternative readings (Fernández Casado 2008 [1972], 89-94).

There is further evidence to suggest that the aqueduct was working and repaired throughout Antiquity, as up to three layers of *opus signinum* are visible inside the *specus* (figure 68). They are below the fifteenth-century repairs made in stone, which they must predate; the upper layers are made in an *opus signinum* coarser than that of the original work, which may indicate late imperial or even post-imperial repairs of the conduit (Jurado Jiménez 2001).

---

[24] *[Imp(eratoris) Nervae Traiani Caes(aris) Aug(usti) Germ(anici) P(ontificis) M(aximi) trib(unicia) pot(estate) II co(n)s(ulis) II patris patriae iussu P(ublius) Mummius Mummianus et P(ublius) Fabius Taurus IIviri muni(cipii) Fl(avii) Segoviensium aquam restituerunt].*

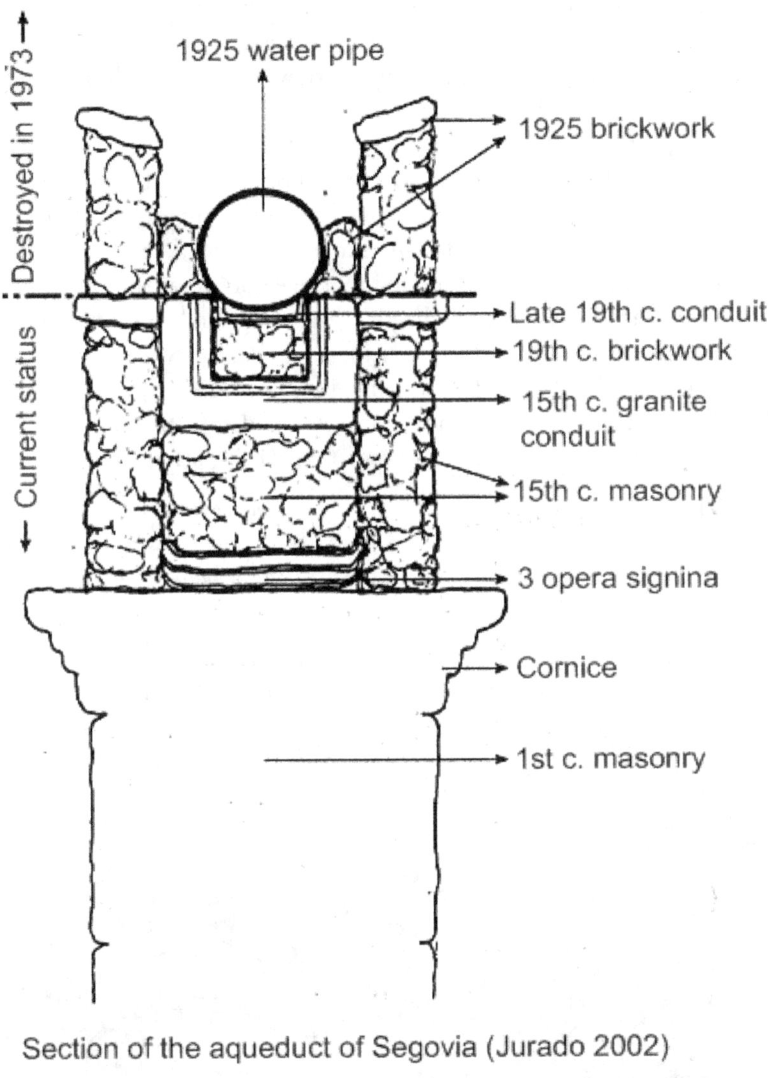

Figure 68: Section across the successive specūs of the aqueduct (Jurado Jiménez 2001).

The important question is whether the aqueduct continued to function after Antiquity, for which there is no clear evidence, but which is an intriguing possibility. Tradition has it (as per sixteenth-century chronicles) that the aqueduct was first damaged in 1071, after a raid

(Gómez de Somorrostro 1820). Regardless of the veracity of this late source, it is still claimed by scholars that this attack caused the destruction of 36 arches, as the twelfth-century city walls include some cornice-stones from the aqueduct, allegedly from the destroyed arches (Fernández Casado 2008 [1972], 65). This is taken as a fact because in the later Middle Ages (in 1484 and 1505) the aqueduct was heavily repaired, as evident from the presence of Gothic arches (figure 69). Nevertheless, the supposed destruction of 1071 until repairs in the late fifteenth century is contradicted by the evidence of the historian and Archbishop of Toledo Rodrigo Ximénez de Rada who wrote in his *Historia de Rebus Hispaniae* (I.8), that the aqueduct that supplied water to the city had been built by King Hispan, a friend of Hercules, and that it still carried water in his day (AD 1243), as the use of the present of the deponent verb *famulor* indicates:

> *Hyspan autem vir industrius, strenuus (...) civitatem etiam iuxta iugum Dorii h(a)edificavit (...) Secobia nuncuptur, ubi aqu(a)eductum construxit, qui miro opera civitati aquarum iniectionibus* ***famulatur****.*
>
> 'Hispan [was] also a hard-working, strong man (...) [who] also built a city next to the Duero mountain range, which is called Segovia, where he built an aqueduct which with wonder **serves** the city with the inflow of waters'

Figure 69: Pointed arches in the aqueduct.

This would hint at some repairs of the aqueduct in the twelfth or thirteenth centuries or no great damage caused in 1071. The aqueduct was certainly extensively repaired in the fifteenth century, by command of Kings John II, Henry IV and Isabella of Castile, and was last repaired in 1868. In 1925 a water pipe was put on top of the fifteenth-century *specus*, and this was only removed during the restoration works of 1973, effectively putting to an end the working life of the aqueduct of Segovia.

Discussion

There is a big gap between the last un-datable, but still Roman, layers of *opus signinum* from the *specus* and the late eleventh century, when the *arcuatio* of the aqueduct is allegedly destroyed first (1071 attack) and repaired later (pointed arches). In between those dates (we do not know exactly for how long) the arches were still standing, which together with the qualities of the water and the information from Ximenez de Rada means that aqueduct was probably still functioning: First of all, because the conduit carried very little water in volume and that water has a near-null content of carbonates (granitic springs), which means that the conduit was not affected by sinter. Secondly, because the structure is very solid and strong: the arches and pillars are still standing. The use of blocks of granite without mortar means that the structure is more robust, and is not as badly damaged by frost or ice (typical in winter in this region) as if it had had mortar joints. Finally, surveys carried out by engineers demonstrate that only in recent years (due to increasing air pollution) is the stone suffering, though the structure is still robust (Jurado Jiménez 2001). The fact that we know nothing about the archaeology of Segovia further obscures the picture we have, but what we know indicates that the aqueduct was certainly standing, and probably working, all throughout the early medieval period.

SEVILLE

The city of Seville (*Hispalis*) was in Late Antiquity one of the most important ones of the Peninsula, as the main harbour that linked the Guadalquivir valley with the Atlantic and Mediterranean, and one of the most active mints of the period (Tarradellas Corominas 2000, Cabrera Tejedor 2016). Perhaps because of this, Seville was one of the main targets of fifth-century Suevic raids and was also chosen (for a short time) as a *sedes regia* in the Visigothic period (Hydat., 115; *Hist. Goth.*, 41–4; *DLH*, III.30). It was also an archbishopric, the most famous bearer of its mitre being Isidore. The city was a stronghold, together with Córdoba, of the old Hispano-Roman elite, and it was here that Hermengild placed his capital during his rebellion against his father in the 580s (Bicl., 574, 585; *Hist. Goth.*, 49; *DLH*, V.38). Hermengild was even commemorated by an inscription (Fernández Martínez and Gómez Pallarés 2001).

Sadly, very little else is known of the city for this period, because the sedimentary action of the Guadalquivir on the one hand and the continuous habitation of its old urban centre have badly damaged the archaeological record of ancient *Hispalis* (García Vargas 2012). Three Visigothic churches are known, although only from the written sources. Two baths are known archaeologically to have existed into the fifth and sixth centuries.

In the early Islamic period, it seems that the churches were still standing, at least in the early period, when one of them was, allegedly, shared with the Muslims as a centre of cult (Manzano Moreno 2006, 256).

The aqueduct

The aqueduct of Seville, popularly known as *los caños de Carmona* ('the pipes of Carmona' referring to the sixteenth-century piped conduit that ran on top of it) is today hardly visible, although its course is very well known (figure 70). Even though its remains impressed locals and visitors for over a thousand years, they did not impress urban developers who, with the consent of the local authorities, pulled the aqueduct down: of the 410-odd arches standing in 1848, 400 of them were still visible in 1901, but only 15 were standing by the 1960s (Fernández Casado 2008 [1972], 168).

These ran for over 4km, in a single row of arches, 3.9m in diameter on average, and built in *opus testaceum*. The three sections standing today show that at points the aqueduct had to run on double arcades, and from old illustrations it is possible to show that some pillars had reinforcing buttresses, which could indicate a late repair. The aqueduct approached the city from the east, parallel to the road to Córdoba (Fernández Casado 2008 [1972], 159-70, González Tascón and Bestué Cardiel 2006, 274).[25]

Figure 70: Standing remains of the aqueduct of Seville.

The water was obtained from an underground aquifer, by means of a 3km tunnel, which has been recently explored and mapped by speleologists. This exploration has revealed that not only is the water still drinkable, but also that the tunnel was carved into the rock with constant dimensions of 2 x 0.85m (figure 71) (GEOS 2010).

Recent archaeological excavations have unearthed a large (450m^2) three-aisled terminal cistern that could be linked to the aqueduct, at the Plaza de la Pescadería (figure 23). Its location within the old Roman city is dubious. It is close to the Alfalfa square, traditional (and unfounded) location of the forum.[26] More recently, however, it has been argued that it was located outside the city walls,

---

[25] Fernández Casado 2008 (1972): 159-70; González and Bestué 2006: 274.

[26] This raises the issue of whether the forum, as in other main river-side sites like Zaragoza, was located facing the harbour.

despite any evidence suggesting the course of the fortifications. The one piece of relevant topographic information, is that the cistern is too low-lying to have fed the old Roman city, traditionally located further south, around the *alcázar*, so it is believed that it was part of a (supposed) branch that supplied the (known) Flavian area which includes the lower-lying site of La Encarnación (García García 2007, García Vargas 2012, 888).

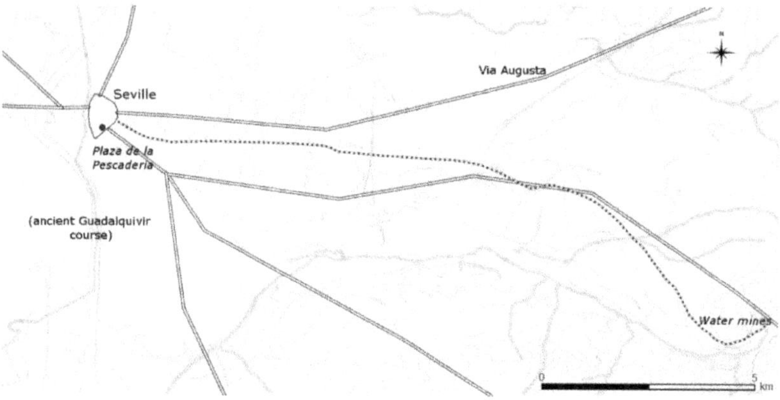

Figure 71: Course of the aqueduct of Seville.

An uncertain chronology

The destruction of most of the aqueduct and the lack of any indication of water consumption prevents the formulation of a certain chronology. The construction of the aqueduct itself can only be ascribed to the second century on the grounds of its construction technique and its similarity to that of the aqueduct of *Italica*, which would fit with the date of the construction of the Plaza de la Pescadería cistern (García García 2007, 136).

The abandonment of the cistern is also used as an indicator of the end of the aqueduct. The earliest ceramic material that has been located in the abandonment layers of the cistern can be dated to the late fifth and early sixth centuries, and these were located in the eastern aisle (where the aqueduct entered the cistern). Another part of the cistern was slightly later turned into a dwelling (?), as the nave was divided by an *opus listatum* wall (figure 72), that has been broadly dated to the Visigothic period. It was then continuously occupied until the eleventh century (García García 2007, García Vargas 2012, 900-5). A sixth century date can be given for the abandonment of the Flavian suburb branch of the aqueduct, but nothing is known about the main aqueduct. However, considering the peak of floods in this period, especially after the fourth century, it would not be surprising to see the aqueduct being badly damaged by the river, which was crossed by an aqueduct bridge.

Figure 72: Visigothic wall, linked to a dwelling structure, inside the Plaza de la Pescadería cistern (García García 2007: fig. 9).

Close to the potential location of the Visigothic cathedral, a possible baptistery has been identified at the site of Patio de Banderas, which has been dated after the fifth century, on the basis of the pottery located in the preparation level of the *opus signinum* floor (Hayes 93A and Keay LIV) (García Vargas 2012, 893, Tarradellas Corominas 2000, 287-7). Despite it having a large *opus signinum* pool, the baptistery did not necessarily require piped water. However, considering the examples of Valencia, Tarragona and Barcelona (where the episcopal complex is closely linked to the Roman water distribution network), a requirement of piped water is possible.

Figure 73: Pipe in the domus of the excavations at La Encarnación (González Acuña 2011, fig. VII 201).

There are two baths known archaeologically which can be dated, according to the material finds, to the fifth and even the sixth centuries: the Cuesta del Rosario baths and the Palacio Arzobispal baths (González Acuña 2011, 487, 492). Apart from these baths, which do not provide any further dating evidence to contradict the dating suggested by the cistern, the excavations at the Plaza de la Encarnación have revealed very interesting domestic fifth-century contexts: in the *domus* known as the Casa de la Columna, a new lead pipe was built into a floor securely-dated to the fifth century (figure 73), and the whole area seems to have maintained the second-century sewers in use (González Acuña 2011, 380-1, Ruiz Bueno 2018b, 155-6).

It is clear that the aqueduct was out of use in the late Islamic period, because according to the Almohad chronicle of Ibn Sahibsala, it was then rebuilt. The chronicle mentions that Abu Jacub Yusuf inaugurated the aqueduct to supply the Alcázar on the 13th of February of 1172. This new conduit was working until the seventeenth century.

> 'He [Abu Jacub Yusuf] is responsible for the water conduction by means of a conduit (*saqiya*) for the supply of the inhabitants of Seville and for its alcázar ... on the old road that leads to this city [Seville] there was an old conduit (*saqiya*) that the soil had covered, so it had been turned into a row of stones by the road, but without anyone knowing what it was... [so the engineer then excavated it, revealing] the remains of a pipe or a conduit which used to supply water to Seville, a work of the first kings of the old Romans [the text continues, to say how the engineer excavated a fountain and later found the source of the aqueduct]' 27

Furthermore, A. Canto (1978, 329) suggests that all the standing remains of the aqueduct have to be dated to this Islamic phase, accounting for inconsistencies in Fernández Casado's description, who based his work on the few standing remains and on old pictures. However this may be, the standing remains do not provide us with any dating evidence, so this whole issue can be mentioned, but not resolved.

[27] Fernández Casado 2008 (1972): 181 [my translation of his Spanish].

Overall, the aqueduct of Seville may well have been in use into the fifth century, and probably ceased to function at some point in the sixth century, when the terminal cistern is definitely out of use, perhaps as a consequence of a flood.

TARRAGONA

Tarragona (figure 20) is a Mediterranean city, located on the north-eastern coast of Spain. It was one of the first cities founded by the Romans outside Italy and it was, from its very early beginnings, the provincial capital of *Hispania Citerior* (later *Hispania Tarraconensis*). The city's layout was greatly modified by Augustus, who spent a year there while he organised the Cantabrian wars (26-25 BCE). After this remodelling, Tarragona was divided into three main areas: an upper city around the temple of the imperial cult and the *concilium provinciae* (a great public space linked to the provincial administration), a lower city, south of the circus which included the old colony's forum and a theatre, and lastly an extra-muros, sea-front harbour suburb – much like Lisbon.

During Late Antiquity, the city remained as the last bastion of imperial power, as it was the *de facto* capital of the *Dioecesis Hispaniarum* after Mérida was lost to the Sueves early in the 420s, and it was the last city to fall to the Goths in 475. Throughout the fifth and into the sixth century, Tarragona maintained its ecclesiastical importance as an archiepiscopal see, but it lost most of its political importance to sites like Zaragoza or Barcelona, especially after 592 when Barcelona took over the taxt responsibility of this region.[28] The city entered into a period of steep decline which continued up to the eighth century, as evidenced by the increasing number of abandoned structures in the lower city and the decline of imports identified from the harbour. Soon after, the city was abandoned by the Muslim garrison and only became again a centre of political relevance in the

[28] According to the *De fisco Barconinensi* mentioned above. An extensive and up-to-date study on Tarragona has been carried out by Meritxell Pérez Martínez (2012).

twelfth century, when it was conquered by the counts of Barcelona (Panzram 2002, 111-20).

The archaeology of the city reflects the history of this waning provincial capital. Only the forum appears to have been maintained functional throughout the fifth century, as evidenced by the lack of encroaching structures in it (Bosch Puche, et al. 2005, 168-71). But beyond that, there is enough evidence to suggest that the city was polarising into two foci, as the population abandoned the lower city and moved either up to the upper one (where the Visigothic episcopal complex was built) or to the harbour suburb, which had been expanding towards the mouth of the Francolí river, around the Christian martyrial complex built in the late Roman period.

Most of the old houses and other buildings were abandoned during the course of the sixth century, and those that were built anew are to be found outside the walls. The circus was then quarried, and the amphitheatre was turned into a centre of Christian cult in 555 with the construction of a church in the arena. Similarly, the finds of the new fifth-century harbour indicate that Tarragona was slowly losing importance in broader commercial networks, and imports cease to appear in the archaeological record by the end of the seventh century, when the few identifiable finds seem to have been of local or regional origin (Macias Solé 2008, Macias Solé and Remolà Vallverdú 2005).

The multiple water supply systems

Tarragona is located both next to a river (the Francolí) and on a karst platform in which the water table is readily accessible, which benefitted supply in the early years of the colony. Despite this, during the early Empire, in the period of urban reorganisation, the city built three new aqueducts.

In the Early Modern period however, when Pons d'Icart (1572) undertook the first studies on the aqueducts of Tarragona, only one conduit was known, the so-called 'bridge of the devil' or *Pont del Diable* (figure 74), which was linked to the few and fragmentary remains known from the city. This first study linked the bridge with two possible sources (the Gayá and the Francolí rivers), although over time the source of the Ferreras aqueduct (its current name) was considered to be the Gayá river alone (an opinion followed by Fernández Casado).[29]

Figure 74: View of the standing remains of the aqueduct of Tarragona.

However, later in the last century this was proved wrong, because the 1972 survey put together the information obtained from field work with a new study of eighteenth- century maps. It was then clear that there were two aqueducts, one obtaining water from the Francolí and another obtaining it from the Gayá river.

Nowadays, we know of the existence of three different aqueducts.

[29] Fernández Casado 2008 [1972]: 29. Fernández Casado's description of the course is completely outdated, but his notes on the remains which were standing when he did his research in the 1940's and 1950's are still important and unparalleled.

The Francolí aqueduct

The water conduit to which the Ferreras aqueduct belongs, as we have said, is now linked to the Francolí river (Sánchez López and Martínez Jiménez 2016, 164-6). Its course is now relatively well known, especially due to the early topographical surveys and maps done in the 1740's as well as to the aforementioned 1972 survey. Its total length is 15km. Its source, at the Hill of Rovel was finally identified in 1989, as previously it was thought to be further down river at Puigdelfí (Remolà Vallverdú and Ruiz de Arbulo 2002, 33).

Figure 75: Remains excavated at the Camino del Ángel (Sánchez Real, Pujante Izquierdo and Pau Bages 1994, lam. Via).

From this area in the hills, the aqueduct continued towards the sea partially excavated in the rock. Remains are known at the sites of Mas Blanquet and visible at the creek of Castellot (Sáenz Ridruejo 1977). From here the aqueduct continued to the area of the Ferreras creek. There we find the arcaded aqueduct: it measures 200m in length and up to 26m in height. It is organised in two rows of voussoir arches (25 arches on the top row and 11 on the bottom one) which measure 5.90m in diameter (20Rf), and are built in ashlar blocks (Fernández Casado 2008 [1972], 31-40). From this point, the aqueduct followed the contour lines until it approached the city from the north, by the slopes of the hill of La Oliva, where it is still visible in Camino del Ángel (figures 75, 76) and the Avenida de Cataluña (and where it used to be visible at Quatre Garrofers), supplying water to the lower city (figure 20) (Macias Solé, Fiz Fernández, et al. 2007, Sánchez Real, Pujante Izquierdo and Pau Bages 1994).

Figure 76: Remains of the aqueduct built adjacent to the walls of the upper city.

The Gayá aqueduct

A second aqueduct reached Tarragona from the Gayá river (figure 77). The course of this aqueduct is not well known outside the city, although its source at Puente de Armentera and its approximate length (45km) are widely accepted (Sánchez López and Martínez Jiménez 2016, 160-3), making it one of the longest aqueducts in Spain. There are few remains known from there down to the city itself, but some remains were excavated at the site of La Secuita, which were partially vaulted (figure 78) (Cortés, Benet and Bermúdez 1989).

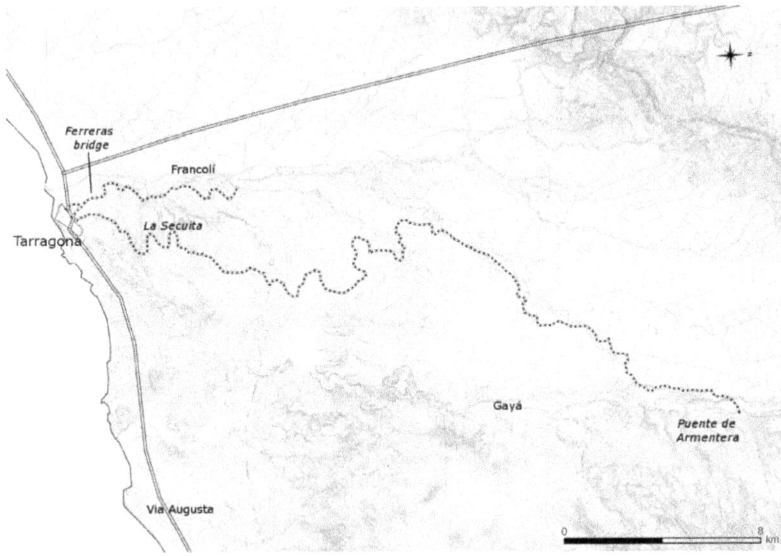

Figure 77: Map with the aqueducts of Tarragona.

Inside Tarragona it is better known, and it seems clear that it fed the upper city. Ten sections of it have been recently excavated, which show that it divided into two conduits just before entering Tarragona at La Oliva hill. The first branch is a vaulted conduit, lined in *opus signinum* and built in mortared rubble, and has been identified in several places of Calle Rovira i Virgili, and went straight towards the circus area. The second branch went towards the upper city, and according to the pillars found at Calle Sant Auguri, it reached this part by means of an inverted siphon. There are some remains still standing at the Campo de Marte at the point where the aqueduct crossed the walls. From this point it seems, according to the distribution of the finds, that most of the water was carried up to the *castellum* located in the archiepiscopal palace, whereas another section of the aqueduct went parallel to the walls, in the via del Imperio Romano (Macias Solé, Fiz Fernández, et al. 2007, items #24, 69, 86, 166, 181, 719, 726, 728, 729, 733). The late inscription that may relate to a fountain can be topographically linked to this aqueduct.

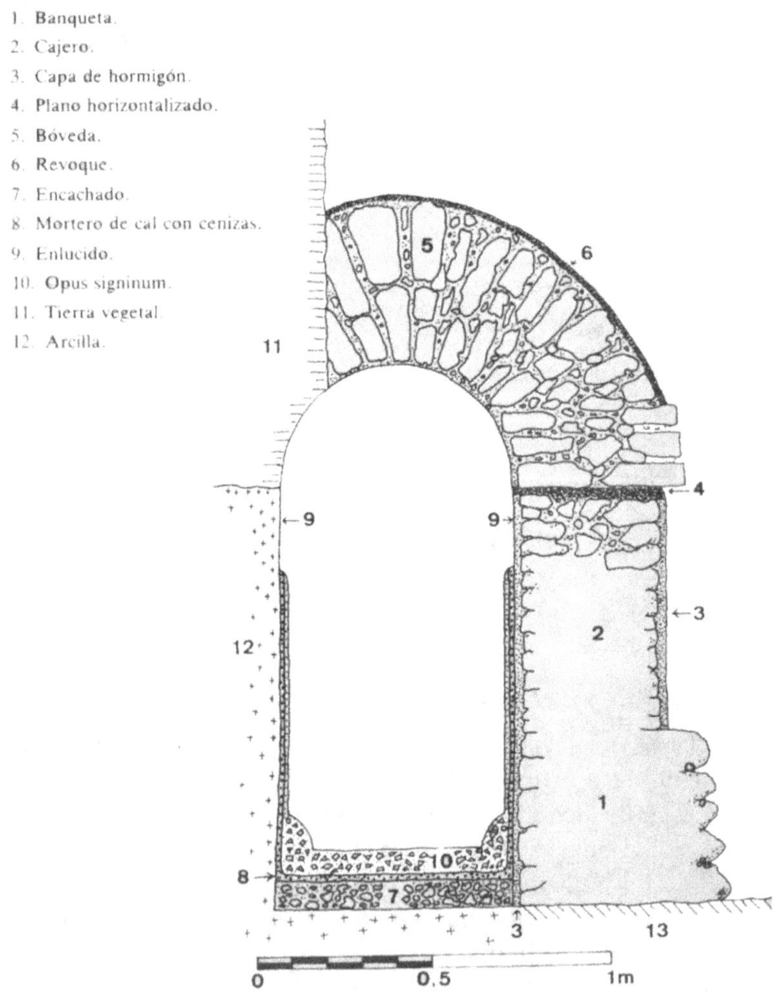

Figure 78: Section across the conduit located at La Secuita (Cortes et al. 1989, fig. 2).

The suburb aqueduct

A third and much debated aqueduct, which in 2003 was still unknown archaeologically, although it was inferred, supposedly fed the harbour suburb and the sea front residential area (Macias Solé and Remolà Vallverdú 2005, 186). Now there are archaeological remains that may indicate that this aqueduct existed: a section was excavated at Calle Dr. Mallafré which because of its altitude (30m a.s.l.) cannot be considered as a suburban branch of the two other aqueducts (which in this area are located 50/60 m a.s.l.).

This suburban aqueduct is thought to come from the Francolí river, diverting water at some point down river from the Francolí aqueduct, in the place known as Sant Pere Sescelades. Even if its course is largely unknown, it is possible that it followed the later medieval water conduit known as the Rec Major, which was used to power the mills of the city, also located in the suburb (Cortés, Benet and Bermúdez 1989, Remolà Vallverdú and Ruiz de Arbulo 2002, 33-4).

The lack of archaeological evidence prevents us from making strong statements on the nature of this aqueduct, beyond the safe assumption that it fed the suburb, perhaps supplying the baths and baptistery of the suburban episcopal complex.

The karst and the cuniculus

The karst platform on which Tarragona is located, which contained an underground lake, made water easily available to the city (Burès, García and Macias Solé 1998). At some places by the harbour suburb the water even naturally sprung out of the soil (Remolà Vallverdú and Ruiz de Arbulo 2002). This would also explain the presence of deep wells all over the lower city (22 are known in the archaeological record).

This underground water seems to have been conducted by means of an aqueduct excavated in the rock, or a *cuniculus* (figure 79), which has been located during some rescue excavations at Calle Gasómetro n. 32, originally discovered in 1849. This conduit, located very close to the *forum coloniae*, is known only for 68 twisting and turning metres. The conduit is cut into the bedrock, and took water from the lake and led it to somewhere by the main harbour. Its cross-section is shaped like an inverted-bottle 2.5m high, with its vault measuring 80cm in width, the passageway 50cm and the conduit itself, at the bottom, just 30cm. The access to this tunnel was through six *spiramina*.

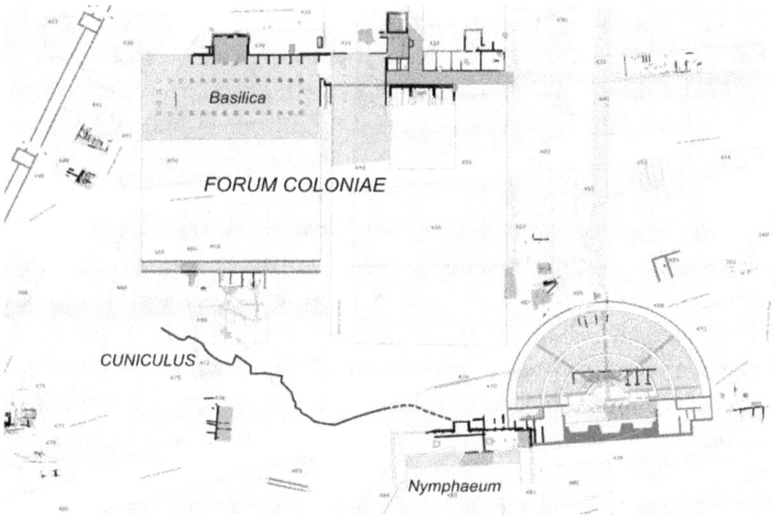

Figure 79: The course of the *cuniculus* in the lower city (based on Macias Solé, Fiz Fernández, et al. 2007, fig. C).

Hardly anything else is known about this conduit, although other similar structures have been located in recent excavations, of which even less is known, but that would indicate that Tarragona's underground waters were commonly used. Some of these conduits seem to be related to the water supply of the theatre *nymphaeum* (Macias Solé, Fiz Fernández, et al. 2007, #453, 464, 469, 498, 511).

The public fountain located at Calle Pere Martell, in the old harbour suburb, seems to have been fed directly by water springing from the ground, and not to any known excavated conduit. This natural spring was enclosed in a pool measuring 15 x 3m with several sculptured spouts. It was monumentalised in the early imperial period with a vaulted roof, which substituted the late republican simple tiled roof and colonnade. In the late Empire it was almost abandoned after its partial collapse, although the pool of the spring was kept accessible and in use (Remolà Vallverdú and Ruiz de Arbulo 2002, 41-63).

The chronology of the water supply

Our uneven knowledge of the aqueducts prevents us from making any clear statements on their chronology. The construction of the Francolí aqueduct has been dated to the Augustan period, on the sole basis of its ashlar construction technique and because Augustus himself was in Tarraco in 26-25 BCE, and was responsible for the great urban development of the city. Out of the excavated sections, no conclusive datable material has been obtained, so all the information necessarily has to come from indirect evidence.

Only the underground conduit, the *cuniculus*, has provided datable material. It seems that the tunnel was partially out of use by the fourth century, because a section of it had by then been turned into a rubbish pit (Burès, García and Macias Solé 1998).

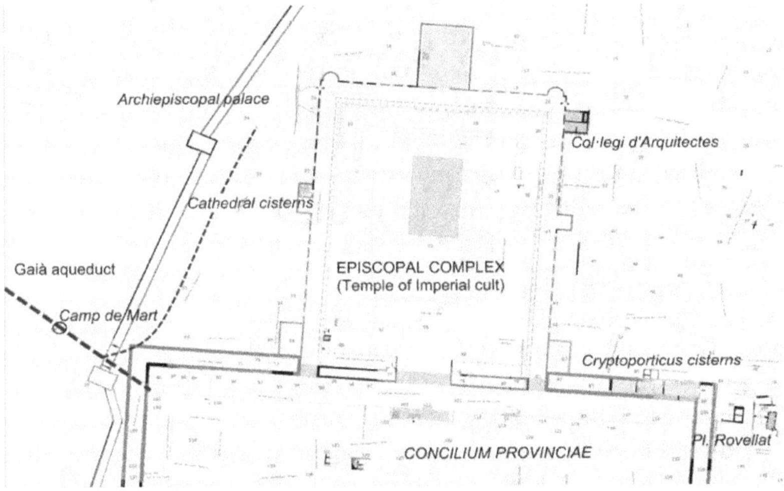

Figure 80: Plan indicating the location of late antique cisterns (based on Macias Solé, Fiz Fernández, et al. 2007, fig. A).

The fifth-century inscription listed above (*CIL* II 4109), as mentioned, might have been linked to a fountain. Because of the location where it was found, at Camino del Cementerio n.7, it could be linked to a fountain of the Gayá aqueduct.

The presence of many late antique cisterns has been used to suggest an end of the piped water supply, and a return to previous systems of water supply. However, this could be an indicator of an unreliable and seasonally variable aqueduct, as explained above. In the case of Tarragona, the large cisterns located at the cathedral could be considered to be one of these terminal cisterns (Bosch Puche, et al. 2005). This cistern has been dated to the Visigothic period according to its construction technique and its location, re-using a wall of the cryptoporticus of the imperial cult *temenos* (which is considered to have been abandoned by the late fifth century). All this fits with the sixth-century transformation of the precinct into the newly-built episcopal complex. The cistern measures 45.83m² and it is covered with a 25cm thick *opus signinum* lining.[30] It is a barrel-vaulted structure, with an overall capacity of 147m³, which is very big. This cistern, furthermore, is very close to the archiepiscopal palace (figure 80), where the *castellum* of the Gayá aqueduct has been identified, and it is very possible that it fed the cistern although this is not certain. Other cisterns in the same area of the city may have also been built to store water coming from the Gayá aqueduct. The *cryptoporticus* cisterns, for instance, were built in the late Empire, after lining in *opus signinum* three rooms, one of which preserves an access stair. It is possible that the *specus* which has been identified in this excavation (which came from the aqueduct itself) was used to fill these cisterns. These were in use up to the end of the fifth or into the early sixth century, when the pottery indicates that they were transformed into a dwelling space (Macias Solé, Fiz Fernández, et al. 2007, #85). Lastly, two more cisterns located in the immediate area have also been dated to the Visigothic or late antique period, although they seem to be too low to be linked in any way to a terminal deposit of the aqueduct. This of course does not imply that they could not still be fed by aqueduct water that was piped from the terminal deposit. The cistern in the Colegio de Arquitectos has been dated to the early Visigothic period (500/525-550), and its size (40m³) may indicate that it was domestic (Aquilué 1993). The date is given by the ceramic material retrieved from other contexts in the same excavation, although there is no specific mention to datable finds in the cistern. Because of the location of these cisterns, it may be possible to link them also to the episcopal complex. Very similar is the cistern of Plaza del Rovel-

[30] Perhaps a sequence of superimposed linings (?).

lat, which is linked to a residential area built in the late fifth and early sixth centuries, according to the excavated pottery (Bosch Puche, et al. 2005).

As for baths (over twenty are known), they were all located either in the lower city and the harbour area or in the western suburb, which means that they were fed by the Francolí aqueduct, the underground *cuniculus*, or the alleged third aqueduct. The main (and only known) public baths, located on the harbour front, were excavated at the Calle San Miguel, and were very well preserved. This made it possible to obtain a full stratigraphic sequence from their origins up to Late Antiquity. The baths themselves were organised with a main axial room and two adjacent wings. They were supplied by the underground conduit and were originally built in the early Empire, being completely refurbished in the third century. It may be that the baths restored by the governor in the early fourth century (*CIL* II² 14, 1004) correspond to these ones. The structure of the baths was turned into a house which was in use during the sixth and seventh centuries, and until the Muslim conquest according to the excavators (Díaz García, García Noguera and Macias Solé 2000, Macias Solé 2004, esp. 58-62, 83-4). Five other private baths continued to function into the late antique period; those of Parque Central and the Francolí Christian suburb seem to have been functioning up to the fifth century (later being remodelled into a baptistery) (García-Entero 2005, 270-2). Two other *balnea*, excavated during the urban development project known as PERI-2, continued in use well into the Visigothic period. That located in plot 22 was a fourth-century domestic *balneum* built in a suburban *domus*, and was in use until the second half of the seventh century, according to the excavated material, which mostly (over 60% of the datable material) consisted of seventh-century coarse wares. The other bath, located in plot 31, was built at the same time as the previous example, but this bath was by the mid-sixth century a burial area. These two baths seem to have been supplied by underground water (Adserías Sans, Pociña López and Remolà Valverdú 2000). The last example is the *balneum* of the *domus* at Calle Apodaca, in the lower city, inside the wall precinct. This bath was built in a house located on top of the main sewer, and was in use until the beginning of the sixth century, according to the excavated material (Macias Solé, Fiz Fernández, et al. 2007, #355).

The Islamic period

The aqueducts of Tarragona, like many others in the Peninsula, were not commented on at all by Classical authors, and it is only in the Islamic period when geographers paid attention to these huge structures. In the case of Tarragona, the aqueduct is mentioned by Ibn Ghálib, a twelfth-century scholar, whose text is lost but has been preserved by the seventeenth-century compiler, al-Makkari.

> '[Ibn Ghálib mentions that the aqueduct] conveyed the water from the sea [sic] to the city by a gentle level, and in the most admirable order, and served to put in motion all the mill-stones in the town, the whole being one of the most solid, magnificent, and best contrived buildings that ever were erected' (Al-Makkari, I.6)

The text must certainly refer to the Ferreras bridge, the Francolí aqueduct, as the other aqueducts have not left remains that could deserve such praise, and it implies that a water conduit in Tarragona was still functioning at the time of the Christian conquest (1116 CE). This text is however very late and problematic, and should not be taken face value, as it may be confusing the Rec Major conduit feeding the mills with Roman remains (Guidi Sánchez 2016, 47). It is difficult to explain also how it uses sea water.

Several modern authors, however, mention that the aqueduct of Tarragona was repaired in the tenth century by caliph Abd al-Rahman III, and even mention that these repairs are 'clearly visible in the current monument'.[31] Yet I have not been able to find a picture of these repairs, a reference to excavated material or a primary Arabic source mentioning them. However, these alleged repairs are not impossible, as we know of Islamic repairs in Córdoba and Valencia, and they would explain the confusing comment of Ibn Ghálib. It seems unlikely it was designed to supply of the city, of which we know very little for the tenth century – or indeed for the entire the Islamic period of Tarragona (Elices Ocón 2017, Guidi Sánchez 2016).

---

[31] Most prominently, Fernández Casado 2008 [1972]): 29, whom the quote belongs to.

Discussion

Even if there is not enough direct evidence from the aqueducts to determine when they ceased to function, there is sufficient indirect evidence, which may point towards the sixth century as the end point of these structures, or at least one of them. There is an unfounded suggestion that one might have been in use and repaired again in the Islamic period, although there is no solid evidence to confirm this.

A possible *post quem* date for the abandonment of the Gayá aqueduct is 468 CE, if the inscription dated to the reign of Anthemius and Leo was really related to a fountain head, which is unclear. The proliferation of sixth century cisterns in the upper city (some of them linked to in-flowing *specūs*) would give another, rather later, *post quem* date. This argument, however, relies on the assumption that the huge cisterns located in the area of the cathedral can be linked to a partially functioning aqueduct and to a means of maintaining a regular water supply, as happened with the terminal cisterns of the Aqua Antoniniana of Rome. This suggests parallels with the continuing aqueducts in the episcopal complexes of Barcelona and Valencia, and may hint at a direct involvement in the water supply systems by the Church.

It should be remembered that one of the branches of the Gayá aqueduct fed the lower city, just like the Francolí aqueduct. Furthermore, in this area we know the baths of Calle Apodaca, which as we have mentioned, were in use until the beginning of the sixth century. This could support the dating proposed for the Gayá aqueduct, unless the baths were linked to the Francolí aqueduct.

As for the two Francolí aqueducts, the main one cannot be certainly linked to any water-consuming structure, and the existence of the secondary one is in doubt, so there is very little that can be said about them. It would seem that the harbour area and the lower city relied on underground water for their supply (as evidenced by the many wells in houses). This can be further confirmed by the fountain located at Calle Pere Martell, which seems to have been in use until the seventh century, when the pool was finally filled with sediment and datable ceramic material. The presence of four private *balnea* in this area, dated up to the fifth and even the seventh century, could indicate that the water supply continued to function, but it seems that water was obtained from the subsoil rather than from pipes.

However, the public baths from Calle San Miguel were in use until the mid-fifth century, whereas the conduit that fed them water (the *cuniculus*) was out of use by the fourth. If these baths were functioning until then, they must have been fed by the Francolí aqueduct, because it fed that area. There is the chance, though, that in that period the baths were supplied from another subterranean source of water of which we know nothing.

Overall, the aqueducts of Tarragona do not seem to have survived beyond the Visigothic period, and the only solid evidence points towards an end during the sixth century for the Gayá aqueduct. Regarding the two other aqueducts, only suppositions based on little solid evidence, point towards a late-fifth abandonment date.

VALENCIA

Valencia was a small coastal town in the province of *Carthaginensis*, located on the south bank of the Turia River. It was a Republican veteran colony, and an important harbour on the route between Tarragona, the Balearics, and Cartagena. It had an early Roman phase of monumentalisation, but apparently destroyed during the course of the third century, and later rebuilt (Ribera Lacomba 2008, 303, Löx 2017). When it was rebuilt, it was fully furnished with new buildings, including a curia and a lavish *nymphaeum* in the forum, which was in turn enclosed by a wall. Later on, in the area of the forum where the *kardo maximus* and the *decumanus maximus* met, a new sewer was built during the fourth century to substitute the old decumanus sewer that had collapsed (Ribera Lacomba 2005, 210-2). The town had a circus too, which was in use until the fifth century. At this point the whole city seems to go through a period of stagnation until the mid-sixth century, thanks to the new commercial dynamics favoured by the Byzantine invasion and especially thanks to the works of Bishop Justinian (Linage Conde 1972).

The presence of a bishopric in Valencia has been dated back to the time of Saint Vincent, who preached and was martyred there.

The presence of a local martyr allowed Valencia to generate a local cult. By the time Justinian was bishop there were several churches in Valencia, as known from two fragmentary inscriptions *ICERV* 279[32] and *CIL* II^{2} 14. 90 = *ICERV* 356[33]. Amongst these, the excavated suburban martyrial centre and monastery are likely candidates. He was also probably responsible for the construction of the cathedral which has been excavated at the site of La Almoina (figure 81).[34] Though only the apse and part of its outer walls have been excavated, it is possible to tell that the cathedral, even if built by the forum, maintained the orientation of the urban grid, and did not block any existing streets. As time passed, this original complex was expanded, and two cross-shaped structures were built in the late sixth century at either side of the apse, this time directly on top of the *kardo maximus.* These two structures have been identified as a mausoleum or funerary chapel for the bishops of Valencia (known as the 'Prison of Saint Vincent') and as a baptistery with a drain leading to the main sewer (Ribera Lacomba 2005, 214-9, Soriano Sánchez 2000).

Still in the forum area, north of the cathedral, during the seventh century new buildings were erected on top of the previously public spaces, which include an apsidal structure (probably a martyrial shrine), a large monumental well (the 'great well') and a polygonal building (called *macellum* by the excavators). These buildings, however, respected the still-standing and still-functioning Roman structures that occupied that part of the forum: the curia and the *nymphaeum* (Ribera Lacomba 2005, 228-33, Ribera Lacomba and Rosselló Mesquida 200, 179-80).

---

[32] *...]Iustinianus caelebs pontifex sacerd[os]/no<v>a te(m)pla co(n)struens vetustaq(ue) rest[aurans]/...*

[33] *constructu(m) r[...cu]urrunt/, fastigium quis[...a]ula/ nempe nam im[...a]nnos/ hoc pro<v>idens [...]ntis/ tertio antist[es...] anno/ robore contri[...]rimbi/ aptantur hi mir[...]sius idem/ ful<v>ida pr(a)etera [...]a prossus/ lammina sub lato [...]mine aur i/ [cu]lmine cu(m) solid[...] quinque [columnis?]*

[34] Here I stand corrected from what I originally said (Martínez Jiménez 2011) by Markus Löx (2017), as the basilica is not built in the forum, but next to it.

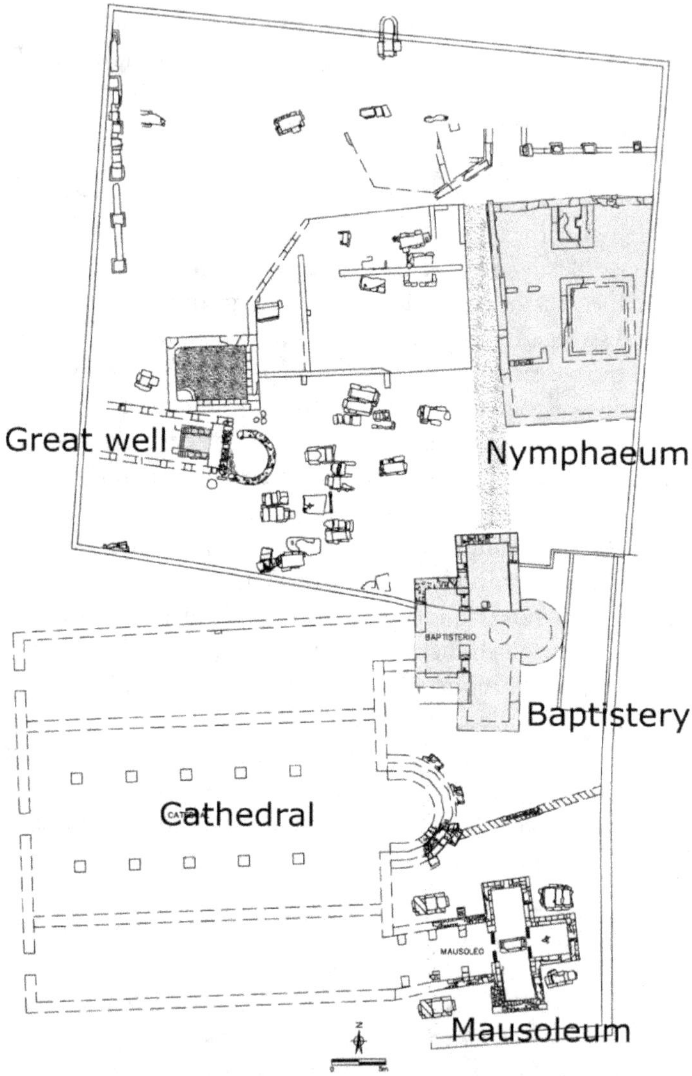

Figure 81: Plan of the Visigothic phases of the excavations at the site of La Almoina (Ribera Lacomba 2008: Fig 8).

Besides the excavations at La Almoina showing the Christian monumentalisation of the forum, little is known from the rest of Valencia. We know, for instance, that the circus was partially reconstructed as a fortification during the last decades of the sixth century, and this has been explained as the result of the full incorporation of Valencia into the Visigothic kingdom during the reign of Liuvigild (Ribera Lacomba 2008, 313). As Valencia was very close to the imperial territories on the south-eastern coast, it became important to take control of Valencia to create a definite border against the Romans. This is further supported by the construction in the late sixth century (according to the excavation material) of the fortification of València la Vella, a 4ha *castrum* ten miles away from the city (Rosselló Mesquida 1996), which further indicates the military presence of the Visigoths in the region. It is certain that by 546 (when a synod was celebrated) Valencia was only loosely controlled by the Visigoths, but during the last part of the sixth century the city was finally included in the Gothic kingdom.

From the later Visigothic to the early Islamic period, the site of La Almoina hardly changed and was only slightly transformed. During the course of the late seventh century a workshop area developed in the northern sector of the old forum. The eighth century is described as the darkest century of the archaeology of Valencia, because very little is known, and what is known can only be ascribed to this century with great difficulty. According to the written sources, the city was destroyed in the year 778 by the troops of emir Abd al-Rahman I, although this destruction has not been identified yet archaeologically. It is not until the ninth century when real changes in the layout can be seen, with the construction of a *hamman* (Islamic bath) in the cross-shaped mausoleum, the erection of new great cisterns and the construction of a new water supply conduit, which is thought to have supplied the emiral citadel (Martí and Pascual 2000).

The aqueduct

The aqueduct of Valencia is not very well known, for two main reasons: the course inside the city is documented in only three sites (one unpublished), and outside the town all the Roman water conduits are considered to be rural watering channels (Martínez Jiménez 2011).[35]

[35] See the Preliminary notes on p. xi.

The rural aqueducts have been surveyed on both sides of the river Turia, and four are known on the north side while five are known on the south side. The north ones cannot be linked to Valencia because their courses go deep into the agricultural region north of the city (Pérez Mínguez 2006).

The five conduits known from the south bank had only been partially studied until very recently. The only major study until 2008 was a short work which dealt only with the sections standing in the municipality of Ribarroja, published by Domingo Fletcher Valls based on an 1849 unpublished survey and his own field work. In his work, Fletcher mentions two different conduits: he identifies one of them as an irrigation channel branching away from the other, the urban aqueduct, which in turn can be identified in nine different sections. The main problem with Fletcher's work is that his chronology is not reliable (he describes the site of València la Vella as a Roman camp), and most if not all of the sections he mentions as existing in 1964 have certainly now disappeared under new urban developments, so it may be impossible to compare the date of these standing sections.

Fletcher proposed a hypothesis for the urban aqueduct: its source may have been a diversion dam located at the Turia river in the village of Villamarchante, located on the right bank (Martí Matías 2001, 24). The aqueduct would then continue its course parallel to the river on its south bank until it appears again at the site of València la Vella, in the village of Ribarroja (16km from Valencia). Fletcher mentions sections standing at the sites of Perpinyanet, Muncholina and Porxinos (all between Villamarchante and València la Vella), and Barranc del Pou, these are the names of creeks (*barrancs*) that flow into the River Turia, and are therefore the logical places to find an aqueduct bridge. It would then continue towards Valencia, appearing again at Manises, assuming that the Islamic bridge located there (and popularly known as *els arquets*, or 'the little arches') is a reconstruction of an original Roman structure (Fletcher Valls 1958).

However, Ignacio Hortelano Uceda published in 2008 an up-to-date study of the aqueducts of the south bank. He identified five different aqueducts, which he classifies as rural watering systems, on the grounds that they are not covered (following Vitruvius' advice), and that they are cut into the soil rather than built out of masonry (Hortelano Uceda 2008). The only parts of these aqueducts which are built on masonry structures and lined with *opus signinum* are those parts when bridges needed to be built. The conduits themselves are very wide (over 1.5m on average) which is also unusual for urban aqueducts in Spain. All of these aqueducts are fed from river diversion dams (which have been preserved), which Hortelano also points to as a sign that they were not intended for urban supply.[36] He argues that they ceased to function in the fifth century.

According to him, these conduits can only be linked to Roman rural settlements, and that they cannot be linked to the one Visigothic settlement known for this period, the palace of Pla de Nadal. However, the Visigoths mention very specifically rural water conduits in their laws (*LV* VIII.4.31: *De furantibus aquas ex decursibus alienis*), and it is evident that there was a pre-existing irrigation network around Valencia before the Muslim invasion (Glick 1970, 190-2). The assumption that river water was not apt for urban supply is, nevertheless, not valid, as there are many examples of uses of river water (from the upper course) for urban supply, including Sagunto, only 30km away from Valencia (Civera Gómez 2008).

Furthermore, Hortelano claims that none of these aqueducts could have possibly continued towards Valencia. Hortelano presents in his paper a model for each of the five aqueducts, in order to estimate their course beyond the point where they are last attested but without any archaeological evidence to support it, and his model predicts that all aqueducts would have turned south, towards the agricultural hinterland of Valencia (known as the Pla de Quart), where he identifies a great number of rural settlements. Aqueduct number 5, which is the one proposed by Fletcher as the one continuing towards the city, also turns south in Hortelano's model, so according to him it could not have fed the city.

Certainly, if any of these canals continued as far as the city, its course is unknown until it reaches Valencia itself, as the whole bank of the river between Manises and Valencia is heavily built over. The

[36] This is probably influenced by Feijoo's thesis.

aqueduct has been located in three different excavations along the ancient road leading west, a continuation of the *decumanus maximus* (figure 82). In the 1930s a section was located outside the provincial gaol, but was later destroyed, although nearby on January 2013 a new section was identified (Machancoses López 2015). Down the road, the excavations at Calle Caballeros and Calle de Quart that took place in the 1990's confirm that the aqueduct continued in a straight line towards the centre of the town, although at some point it turned south, towards the *Porta Sucronensis* (southern gate) (Guillén Fernández 2003, Herreros Hernández 1996, Viñes Pérez 1997), where the *castellum aquarum* is supposedly located according to a fragmentary inscription recorded by Schulten (Martí Matías 2001, 51), which reads:

> *...]um qui aquam trahi[t... / ...]m a Porta Sucronens[i... / ...] empturum V kal(endas) maias.*
>
> '... who brought the aqueduct... / ... from the Sucronensis gate... / ... he will buy (?). [Given] on the fifth day before the calends of May [26th of April]'

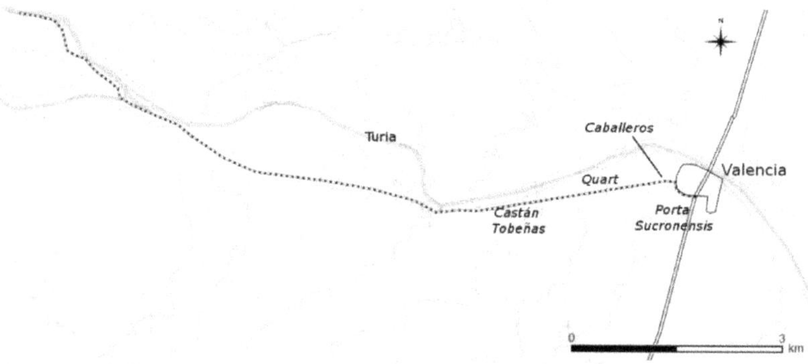

Figure 82: Plan of the reconstructed course of the aqueduct.

It would then be distributed from there to the rest of the town. This included the supply to the monumental forum, with its *nymphaeum* and, later on, its baptistery. Close to the southern gate, the *Porta Sucronensis*, the 1993 excavations at Calle Avellanas unearthed a couple of square bases, which have been identified as two possible aqueduct pillars (Algarra Pardo, Matamoros and Viñes Pérez 1993, 14), which could confirm the text of the inscription although their dating is doubtful.

The aqueduct identified at these sites (Castán Tobeñas, Caballeros and Quart) forms a straight line which runs parallel to the main road that linked Valencia with Segóbriga and Toledo, and as Roman aqueducts normally ran parallel to main roads (which made repairs and maintenance easier) this adds value to the argument that links the urban aqueduct with a source on the right (south) bank of the Turia, even if this may not be one those identified by Hortelano.

Chronology and dating evidence

There is very little evidence to date the aqueduct of Valencia *per se*, and the available evidence is difficult to interpret. It is clear that it had stopped functioning between the seventh and the eleventh centuries, but the continuity between the fourth and fifth centuries is doubtful (Martínez Jiménez 2011).

On the one hand, it is possible to say that the construction in the late sixth century of a large fort outside of Valencia (the site of València la Vella) in order to control the town and its territory, would make more sense if Hortelano's aqueduct number 5 was still functioning, so it could supply water to the garrison there (Martí Matías 2001). The source of this aqueduct is about a kilometre away from the site, so it could have been easily maintained, if at all needed.

On the other hand, there is excavated material, obtained from the inside of the *specus* of the Roman aqueduct located at Calle de Quart, which ranges from the Roman period to the eleventh century (Martí and Pascual 2000, 513). The *specus* located there was totally filled with clay sediments (figure 83). There was plenty of material in these sediments, securely dated between the Roman and the Islamic periods (Herreros Hernández 1996, 15-6). The bottom-most context sitting on top of the clay *specus* had two heavily eroded shards of Hayes 32B and 196, which are generally dated to the third century CE. The layers above it (1144-1146) were all much thinner and all contained eleventh-century pottery and debris. This means that at least until the eleventh century the aqueduct was still carrying water, when it was completely packed with clay, becoming useless.

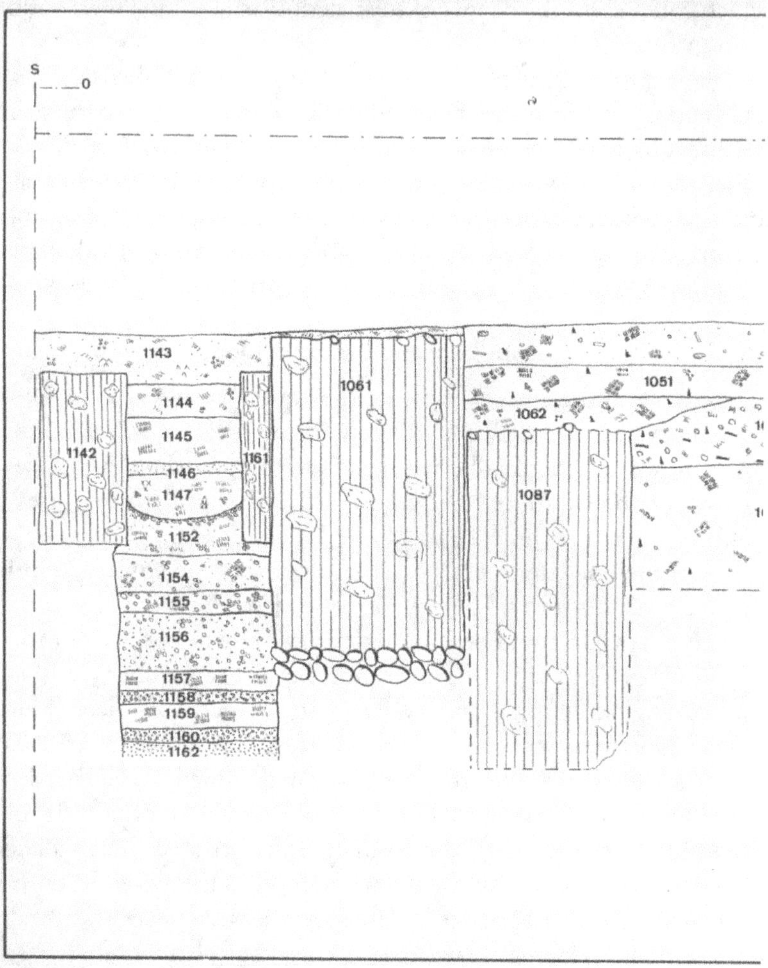

Figure 83: Section drawing of the aqueduct identified at the Calle Caballeros (Herreros Hernández 1996: plan 10a).

The nature of these sediments in context 1147 suggest an alluvial formation of the layer. However, the chronological issues are more intriguing. Context 1147 must have formed after the third century due to the *t.p.q.* given by the two ARS forms. This would seemingly match the third-century destruction of Valencia. However, the layer is very thick, which implies that it carried water for a long period of time, perhaps malfunctioning: the presence of post-third century water-consuming infrastructure in the forum may imply that the aqueduct was still in use then. The presence of Islamic material in the contexts immediately on top of 1147, together with the fact that 1147 was then compacted by a gravel layer, 1146 (which served as the base for the new Islamic phase of the aqueduct), may indicate that in the Islamic period the conduit was cleaned and put back into use. This would mean that the contexts between the third and the eleventh century had been removed.

Beyond the data obtained from the aqueduct itself, it is possible to obtain information on the continuity of the water-supply from water-related structures of Valencia, namely the baptistery, the *nymphaeum*, the old basilica, the sewers, the great well, and the new Islamic conduit.

The baptistery, as discussed above, was first built during the sixth century, added to the west end of the cathedral, north of the apse. It was first identified in the 2002 excavation, but it has not been fully excavated due to the nature of the site.[37] Nevertheless, the excavation has shown that it had a pool with a drain, which could have led the blessed water out of the baptistery into a basin on the street, where the faithful could collect it (Ribera Lacomba 2008, 306). It is not clear where the water for the baptistery came from, but it is not impossible that being so close to the *castellum*, it was fed with water from the aqueduct.

The so-called *nymphaeum* was built during the early years of the colony, although back then it seems to have been simply a sacred pool, linked to some water cults (Albiach Delscals, Espí Pérez and Ribera Lacomba 2009, Álvarez García and Ribera Lacomba 2003). The imperial-phase *nymphaeum* is divided into two areas, although only half of the front one has been excavated. It is all enclosed in an *opus vittatum* precinct measuring 20 x 11m. The front part, which

[37] This has made the identification of this structure as a baptistery quite dubious, but it is a structure with hydraulic infrastructure (*opus signinum* floor with a sewer) located by the apse of a cathedral. A baptistery is the most sensible guess. Of course it could be something else (a bath?), but the point is that: (a) it is a water-related structure, and (b), it can be securely dated to the late sixth-century.

faces north, includes two twin pools and two water fountains fed by pipes; the pools are lined with a layer of *opus signinum* 30cm thick. Although it was originally a sacred well, the presence of lead pipes underneath the first-to-third century pavements suggests that it was fed by the aqueduct, especially as the *Porta Sucronensis* is only 100m away. The main pool is now below an Islamic phase, so it is not possible to know when the original pool went out of use, although preliminary material suggests an eleventh-century abandonment (figure 84). The *nymphaeum* was monumentalised in the fourth or fifth century, refurbished and repaved with new tiles, although some of the walls were pulled down in the fifth century according to the pottery retrieved from the trenches. This, though, did not mean that the *nymphaeum* went out of use, as the fountain was still standing and accessible during the sixth and seventh centuries, as a new marble slab pavement indicates. During the sixth century some obscure, and very difficult to interpret, timber structure (evidenced by the presence of postholes) was erected in the precinct of the *nymphaeum*, but it did not block access to the pool.

In the area between the curia and the cathedral, facing the *kardo maximus,* once stood the old Roman basilica, which in the fourth century was re-modelled to serve a different purpose (Álvarez García, Ballester Martínez, et al. 2005). The old building was divided into several rooms around a courtyard with a drain and a large water basin (4.4 x 9m) lined in *opus signinum.* One of the rooms had a press, and together with the material retrieved from the building (including fish bones, grape seeds, charred plants and fish and oil amphorae) indicates that the basilica had been turned into a food-processing building. The basin, the drains and the vat indicate that the building was consuming water until its final abandonment in the second half of the fifth century. The date is given by the amphorae (which include Keay IV, XXXV, XIII, XXIII and XIX) and the coins found in the abandonment layers.

The 'great well' was built in the north area of the forum, in front of the apsidal structure thought to be a martyrial shrine, and both were probably part of the same construction effort, which might point towards a liturgical function (Vizcaíno Sánchez 2009, 365). The well was lined with reused slabs and ashlar blocks during the

Figure 84: Photograph of the various pools of the nymphaeum of La Almoina (photograph courtesy of the SIAM, Valencia).

seventh century, according to the stratigraphy. The foundation trench includes Hayes 58 and 59 (dated to the fourth century), as well as 91, 99 and 104 (dated 6th-7th c.), together with amphorae Keay LXI, LXII and LXV, which give a similar date (Álvarez García and Ribera Lacomba 2003, 108-9). The presence of this well could be an indicator of a non-functioning aqueduct by the seventh century, built in order to provide water to the core of the episcopal complex and the hub of the city. It was finally abandoned in the eleventh century, according to the material obtained from inside the well, which includes Andalusian large globular *pithoi*, with weathered walls and remains of leather around the neck (showing how water was drawn from the inside) (Martí and Pascual 2000). The well must have supplied water at this stage to the workshops that had been built in the north Almoina sector during the course of the tenth century.

Lastly, the new Islamic water conduit provides us with further evidence to suggest that the aqueduct was operational during the middle Islamic period. This new underground water conduit was constructed in the tenth century (according to the material retrieved from the foundation trench) most probably to supply water to the Islamic citadel, and maybe indirectly from there to the *hammam* (Albiach Descal, Ballester Martínez and Rosselló Mesquida 1999, 46-8). Even if it is impossible to indicate with certainty where it came from or where it went to (because it is only known from this excavation), it probably was a branch diverting from the original Roman aqueduct (figure 26) heading towards the palace mentioned earlier, located to the east of the site of La Almoina, because it crosses the excavated site from west to east. What seems clear is that the conduit was built out of reused material (ashlar blocks, column drums, millstones) and mortared rubble, and that it crossed the old forum without supplying water to the workshops that had been constructed there. This suggests that the water supply system in this period was not openly public and was limited to the palace supply, and maybe perhaps the *hammam* (which it is not clear if it was private or public).

The *hammam* was built in the mausoleum at some point after the construction of the Almoina conduit (9th c.), which allegedly fed the baths, and was in use until the eleventh century. This date is given by the presence of green-and-brown Andalusian pottery in the abandonment layers. The old mausoleum was subdivided with walls, and its floor was raised, in order to warm the rooms. A furnace was added outside the apse, where the hot room was built (Martí and Pascual 2000).

Islamic repairs or evidence for Gothic continuity?

With the available archaeological evidence, it seems clear that the aqueduct was functioning until the eleventh century, with a gap in the evidence between the seventh and the ninth, and a degree of uncertainty between the fourth and the seventh. This chronology is built by the several *t.p.q.*'s and *t.a.q.*'s given by the dated material.

The *post quem* dates include the reconstruction of the *nymphaeum* (3rd-4th c.), the food-processing building of the forum (4th-early 5th c.), the probable use of running water in the baptistery (late 6th c.) and the public accessibility to the pool of the *nymphaeum*, and its late repairs (6th c.). The *ante quem* dates (excluding the great well) all turn around the eleventh century: the destruction of the *nymphaeum*, the final blocking of the Roman aqueduct at Quart, and the end of the Islamic water conduit. It is the evidence for the ninth century (the *nymphaeum* cisterns and the construction of the Almoina conduit) that generates most doubts.

The new constructions of the ninth century clearly point towards a renewal in hydraulic engineering, probably linked to a new urban period of revival, which could indicate a reconstruction of the aqueduct, which had been out of use probably since the sixth/seventh century. However, the evidence can also be interpreted in a different way, as the repairs of the ninth century removed all late antique layers from the conduit, so it could have been the case that water flowed into and beyond the Visigothic period.

ZARAGOZA

The city of Zaragoza was one of the most important cities in the Visigothic kingdom, as it was strategically located between Toledo and Gaul. Founded as an Augustan colony (its Roman name is *Caesaraugusta*) on top of a previous Iron-age settlement, the city is located at the right (south) bank of the Ebro, which was crossed by a now disappeared stone bridge (figure 85). The city still preserves today most of its Roman street-grid, as well as its impressive third-century walls.

In the early seventh century the city was a centre of learning, as the letters of Bishop Braulio (who kept a long correspondence with Isidore of Seville, and mentions the exchange and writing of many books) indicate. It was also one of the main mints, and its territory included the fertile valley of the Ebro river. Its prosperity and importance is also reflected in the several attacks known on the city during the fifth, sixth and seventh centuries, the most important being those of the *bagaudae* (Hydat., 133-4), those of the Frankish kings in 541 (who besieged the city for 49 days until, impressed by the power of St. Vincent, the Franks left; *DLH* III.29), and then in the 620s by the Basques (*Hist. Goth.*, 63).

Figure 85: Bridge of Zaragoza.

The little known aqueduct

Zaragoza is the first city in Spain which we know had a water supply system (ignoring the 'Iberian' aqueduct of Sagunto), as mentioned in a law of the first century BCE, known as the *tabula Contrebiensis*. In this inscription Rome acts as a mediator between the local Celtiberian towns of *Salduie* (the pre-Augustan name of Zaragoza) and *Contrebia*, because both claimed the use of an irrigation channel. This may also have been a water supply system to Salduie itself, if this channel can be identified with a later medieval one which ran into the city (Abadía Doñaque 1995). Despite the inscription, which is very important because it shows the imposition of Roman water law in the Iberian Peninsula, the conduit referred to is hardly known, and it does not seem to have been a proper aqueduct, monumental or otherwise.

It has been traditionally held that the aqueduct was built in the Claudian period, and that it brought water from the Gállego river, a main tributary of the Ebro on its left (north) side. This is, however, only based on the existence of two pieces of evidence. The first one is the conduit known as the acequia de Urdán, a 500m long rock-cut conduit coming from the north of Zaragoza. The second are the lead pipes located at the bed of the Ebro river in 1804 and 1805 (González Tascón, Vázquez de la Cueva and Ramírez Sádaba 1994). They were recorded and published then, but during the Peninsular War (1808-1814), during which Zaragoza was besieged on three occasions, the pipes seem to have been used to make bullets to fight the French. The importance of these pipes resides not only in the inscriptions they bore (*CIL* II 2992), but also in the fact that they indicate that the aqueduct crossed the river by means of an inverted siphon along the bridge, that it entered the city through its north gate, and that its *castellum* may well have been in the forum area, which is just adjacent to the north gate that faced the bridge (Abadía Doñaque 1995, Bestué Cardiel and González Tascón 2006).

This course of the aqueduct is highly hypothetical, because it is only sustained by the existence of the water-pipes, and it would imply perhaps an unnecessarily long aqueduct if water dams existed on the south banks of the Ebro. Another much more plausible (and more recent) hypothesis is that the aqueduct was diverted from the Huerva river (a tributary to the Ebro on its right side, where Zaragoza stands), and that the pipes actually fed the suburb on the other side of the river. This hypothesis is not only based on various facts, which include the large terminal (?) cisterns located east of the city and the presence of several Roman dams on the Huerva, especially at the site of Las Adulas, which has been suggested as the *caput* of the aqueduct (Escudero Escudero and Galve Izquierdo 2011).

Dating evidence

Given the lack of information about the aqueduct, all the dating evidence has to be obtained from indirect sources. The one certain piece of chronological information is that in the Islamic period there is no mention to any water-conduit, and the palace seems to have been fed by a huge water wheel.

Concerning the continuity of Roman baths, there seem to be two phases of abandonment of baths: one in the third century (when the private baths of Calle Ossau and those of Calle Gavín y Sepulcro were abandoned) and another in the fifth (García-Entero 2005, 283-4), like those at Calle Carrillo, the baths at Calle Universidad, and the suburban baths of Calle Alfonso V (Beltrán Lloris 2007, Casabona Sebastián and Delgado Ceamanos 1991). The two public baths of Zaragoza were abandoned in the fifth century: those opposite to the bridge earlier in the century than those at Calle Santa Marta, which were in use until the mid-fifth (García-Entero 2005, 281). However, of the latter only the hot rooms are known to have been abandoned for certain, since the cold ones could not be properly excavated (as they were beyond the excavation limits): considering that hot rooms were generally the first ones to be abandoned, there is a small possibility that the baths continued to function during the second half of the fifth century.

Finally, most of the evidence points towards a functioning aqueduct into the early fifth century, probably up to the first decades of the second half (Santa Marta baths), as after this there is no evidence to prove that the aqueduct was functioning. Any standing remains of the aqueduct were destroyed with the bridge in the ninth century, or unknown to contemporaries in the Islamic period, who used water from the Ebro river for their needs.

Appendix: Aqueducts and Urbanism in Post-Roman *Tingitania*

Tingitania, the westernmost Roman province of Africa was included in the Empire during the reign of Claudius, even if many cities were granted civic privileges already in the time of Augustus. The province was an underdeveloped corner of the Empire, largely disconnected from mainland Africa, and with much closer maritime links with *Hispania*. The number of cities with aqueducts in *Tingitania* is much lower than in *Hispania*, and archaeologically we know very little about them. Trying to include *Tingitania* and *Hispania* in the same parallel discussion on aqueduct and urbanism continuity would prove an unfair comparison, simply because of the proportional volume of data. For this reason, I have decided to apply the same methodology and approach to the cities of *Tingitania* and to add the conclusions as this appendix in which to develop its own narrative separately.

Geographically, *Tingitania*'s main features are the Gharb, a fertile coastal plain, and the mountains that border it: The Atlas Mountains at the SE and the Rif at the NE. It is crossed by three main rivers, from south to north the Wadi Sala, the Sebu (which in Antiquity was navigable), and the Lucus. The mountainous areas of the Atlas and the Rif towards the east were not heavily populated, and remained largely as tribal territories into the Islamic period. This eminently Atlantic position gives the area a milder Mediterranean climate, with rains concentrated during the autumn.[1]

Chronologically, in Late Antiquity the province was divided into a coastal area between Ceuta and Sala, which remained under Imperial control, and the broader hinterland, which was abandoned by the administration and the army. The late antique period thus is a contrast between the late Roman phases of *Zilil*, *Lixus*, Tangiers and Ceuta, and the sub-Roman or post-Roman phases of the cities beyond. In 429 the Vandals passed through the region on their way to Carthage, but never seem to have had any strong claims on *Tingitania*. Nor did the Byzantines after Justinian's campaign, although the coastal cities appear to have been included in the larger Mediterranean trading links. The region was conquered in the decade of 700 by the Umayyad armies, but until then the cities acted largely as independent polities, which could be termed Berber, or Mauri. With the Umayyad invasion the area was included in the Islamic sphere of influence, and in the period of the 780s-820s, a local Idrisid dynasty established its own independent kingdom.

THE AQUEDUCTS

There were various attempts at identifying water supply conduits during the colonial period, both in the French and Spanish protectorates. These studies never went beyond mere identifications or suggestions based on inscriptions. Most of the aqueducts in *Tingitania* remain to this day hardly known or identified (Pons Pujol and Lagóstena Barrios 2011, Mueden 2010).

Of the known aqueducts, six were certainly urban (figure 86; *Banasa*, Ceuta, *Lixus*, *Sala*, *Volubilis* and *Zilil*), while two others are very doubtful (Tangiers and *Thamusida*). Besides these urban ones, there were military camps with baths complexes (*Tamuda*, *Tocolosida*, Aïn-el-Hammam) fed by aqueduct supplies.

- Of the conduits of the forts, we can only claim that they were in use during their active occupation, which does not go into the

[1] Rabat, located in the Gharb coastal plain, is actually wetter than Cambridge.

fifth century (Bernal Casasola, Raissouni, et al. 2012, Bernal Casasola 2018).[2]

- At *Lixus*, the aqueduct has been identified in surveys outside the city (El Khatib-Boujibar 1992). While the local geology does not favour the digging of wells, cisterns appear to have been the most common way of water supply, with and without the aqueduct. It could appear at first that the aqueduct might have continued into the sixth century, based on the continuity to this date of the J baths and some of the *cetariae*, but the late chronology of these baths is very debatable (Lenoir 1992) and the factories were fed by rainwater cisterns (Sánchez López 2018).
- The aqueduct of Tangiers is mentioned in later Islamic sources, but there are no known remains in the city or its surroundings of a conduit, baths, sewers, or garum factories which could give us any indication (Elbl 2013, 159, Mueden 2010, 38, Pons Pujol and Lagóstena Barrios 2011). It may not have existed for all we know.
- At the sites of *Banasa* (Arharbi and Lenoir 2006) and *Thamusida* (Mueden 2010, 182) we have baths and workshops, and hints towards aqueducts, but the early abandonment of the city, prevents us from making further claims.
- At *Sala*, there is evidence for an aqueduct. The forum *nymphaeum* is, in my opinion, misleading, because is still full of water today and suggests that it taps directly to the water table without a conduit (Ammar 2008). The presence of large lead main along the main street and the presence of baths make its existence more likely (Mueden 2010, 211). Considering the main phase of monumentalisation of the city, we know the aqueduct was in use during the second century, but for Late Antiquity hardly anything is known about the site, or its aqueduct.
- The aqueduct of *Volubilis* is securely dated as out of use to the fourth century (Étienne 1960), although for the city itself we have more data about its post-Roman evolution (Fentress and Limane 2010, Fentress 2017) .

---

[2] Although in the case of Tamuda it might have gone into the 420s, although it is impossible to be more precise in the dating.

- The abandonment of the aqueduct of *Zilil* can be dated to the fourth century, while the city appears to have been destroyed by a fire (and then largely abandoned) in the early decades of the fifth century (Lenoir 2005).

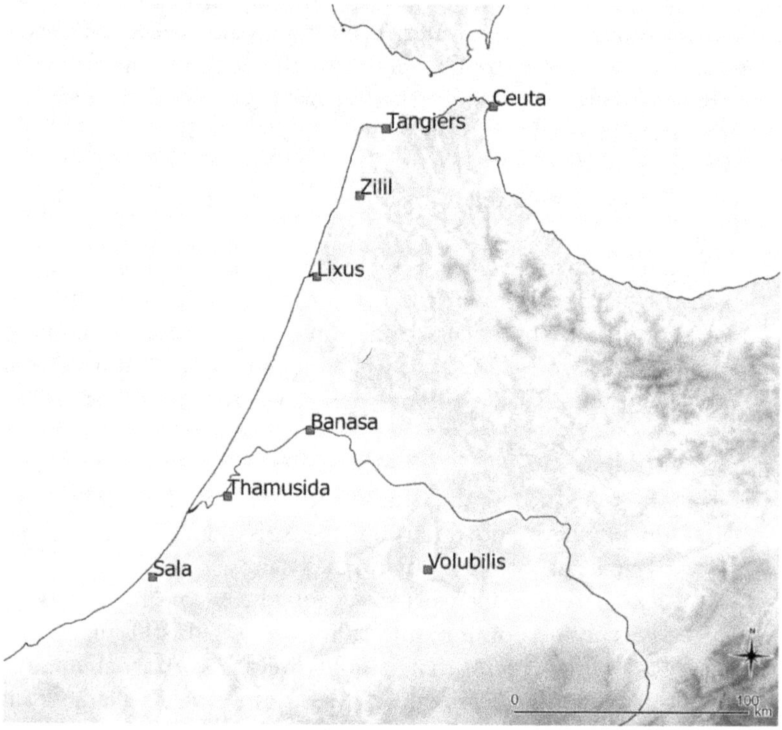

Figure 86: Map showing the location of the urban aqueducts of Tingitania. Note that the Roman frontier was pulled back to a coastal strip down to Sala in the late third century

Regarding the chronologies, only one, the aqueduct of Ceuta, appears to continue into Late Antiquity with any degree of certainty.

Ceuta

The city of Ceuta is located on a small peninsula linked by an isthmus to the mainland. Its connection with the Iberian Peninsula has always been very strong, stronger even than with the rest of North Africa. Furthermore, Tangiers, the closest episcopal see (which probably Ceuta belonged to) was included in the Hispanic Church still in the seventh century (Gozalbes Cravioto 2000). The site does not seem to have been a Roman city of importance, although there was a strong fortified settlement (*oppidum*) on the isthmus and the slopes of the peninsula, which lived, as many other settlements of the region, off the production of fish products, taking advantage of the migratory movements of fish (tuna, mackerel) from the Mediterranean into the Atlantic and vice versa.

The settlement seems to have been deeply affected by the crossing of the Vandals in the late 420s, when one of the necropoleis and some domestic structures were finally abandoned. Or at least this is what the little known excavated remains of the isthmus indicate (Bernal Casasola 2008). The Vandals do not seem to have settled there, as told by Procopius in his account of the Byzantine conquest of 534 CE (*Ktis.*, VI.7.14; Cf. *CIust* I.27). Justinian's planned invasion of Ceuta was a way to control events in the Visigothic and Frankish kingdoms, but the city indubitably was also an important landmark (the Pillars of Hercules) and a stronghold for the later invasion of Spain in the 550s (Vizcaíno Sánchez 2009, 130-2). The imperial occupation of Ceuta meant a new period of urban development, partially planned by the central government (Procopius mentions the restoration of the basilica of the Theotokos and of the walls: *Ktis.*, VI.7.14), but also by the local authorities (the construction of a new street leading to the harbour may be related to this) and private initiative (as indicated by the construction of new houses). Ceuta remained in imperial hands until the 680s (despite Visigothic attacks on the city; *Hist. Goth.*, 42), and was then taken over by the Visigoths. According to legend, Julian, the Visigothic count in 711 was responsible for the crossing of the Umayyad troops into Spain, causing the end of the kingdom (Ibn Abd al-Hakam 42; *Cont. Hisp.*, 68; *Hist. Lang.*, VI.46).

The archaeology of the city is fairly well known, especially thanks to the many archaeological interventions carried out in the last years (Vizcaíno Sánchez 2009, 130-3, Bernal Casasola 2018, 114). The most impressive monuments for the late antique period include a funerary enclosure (the so-called 'Byzantine basilica') (Bernal Casasola and Pérez Rivera 2000, 123-31), a new street with new housing (Bernal Casasola, Pérez Rivera and Carvajal Montero, et al. 2005,

438-41), all linked to an abandoned set of *cetariae* at the Paseo de las Palmeras site.

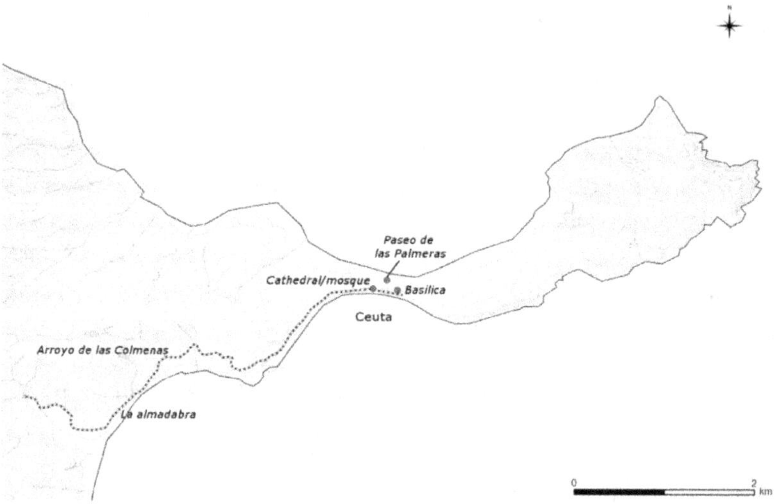

Figure 87: Reconstructed course of the aqueduct of Ceuta (Posac Mon 1977, fig. 1).

Two factors favoured the construction of an aqueduct in Ceuta: firstly, the lack of easily accessible water in the settled area and, secondly, the production of fish sauces. The aqueduct is today completely lost, and its course has to be reconstructed from the evidence noted down earlier in the twentieth century and in the description of the city of Ceuta by Abu Abdullah al-Bakri (1014-1094), who also recorded the Roman aqueduct of Tangiers.

The aqueduct is described in full by Carlos Posac Mon (1977). The source of the conduit is located at the Arroyo de las Colmenas, known in al-Bakri's text as the Awiyāt river, from which it went, following the contour lines for 3km up to the southern bay of Ceuta, where the Roman settlement was located (figure 87). The source says that this conduit fed the Main Mosque, near where the Cathedral is located today. Archaeologically, two sections are known, although they have now disappeared. Soon after the aqueduct left the Arroyo de las Colmenas, the aqueduct had to cross over a valley, which was done by means of a 15m-tall, triple-ached bridge (figure 88), and further downstream, a section was located in 1901, known as the *specus* of La Almadraba. The latter remains were destroyed at the time of the Spanish Protectorate (1913-1956), when a new road was built between Ceuta and Tétouan, which followed the same contour line as the aqueduct.

Figure 88: Last standing remains of the aqueduct of Ceuta (Posac Mon 1977, lám I).

With the aqueduct largely gone, and without any extensive archaeological study, all the dates are hypothetical. The aqueduct is supposed to be Roman in origin and the construction technique suggests this is so. It has to pre-date the eleventh century, when al-Bakri describes it, and possibly has to be dated before the second century, when the fish factories at Calle Gómez Marcelo were established (Bernal Casasola, Pérez Rivera and Carvajal Montero, et al. 2005, 436, Pons Pujol and Lagóstena Barrios 2011, 535), even if al-Bakri indicates that it was built in the 710s by Count Julian, who is a figure deeply shrouded by mystery.

The date of the fish factories is key, as I would suggest that the settlement, not being an important town in the Roman period, built the aqueduct for this only purpose. Thus, the second century CE could be an acceptable date for its construction. As far as the abandonment is concerned, the evidence is increasingly better (Bernal

Casasola 2018, 108). The fish factories known from Ceuta (Paseo de las Palmeras, Gómez Marcelo, and Queipo de Llano sites) were in use until mid-fifth century, perhaps ito the early sixth. A similar sequence is known from the purple dye factory recently dug at the site of Plaza de África. Furthermore, the production of amphorae in Ceuta (types Keay XIX A and B) has been confirmed by means of petrological and chemical analyses, and these typologies are dated between the third and the mid fifth century (Bernal Casasola 1996). This type of amphora, furthermore, is linked to the transport of fish products (remains of mackerel have been identified in several amphorae of this type in Ceuta), so it is reasonable to assume that the production of fish products continued beyond the abandonment of our only excavated factory. Both fish factories as well as pottery productions require large amounts of fresh water, which are not easily available in Ceuta, unless water was brought by means of an aqueduct. This would make a fifth-century continuity of the aqueduct much probable.

After the end of amphora productions, there is no known element of the urbanism of Ceuta that would have required a supply of aqueduct water. Once Ceuta became a Byzantine fortified enclave and a main harbour (with an increase in the number of imported eastern materials, indicating an active trade), it could be argued that a supply of water was needed for the garrison or for the navy, but this is highly hypothetical. Besides, it would be surprising if Procopius failed to mention any repair of the aqueduct in a city with such a symbolic importance, which was considered the 'threshold of the Empire', so it is doubtful that the aqueduct was repaired by the imperial administration, although this does not exclude the tiny possibility that it was still working.

One last thing should be noted, and it is that in al-Bakri's story the aqueduct reached the mosque, which may indicate that, just as happened in Córdoba, an old water conduit was repaired or diverted to feed the fountains of the mosque at some point before the eleventh century (probably, as in the case of Córdoba, in the ninth or tenth centuries). This is again parallel to his description of the alleged aqueduct of Tangiers (Gozalbes Cravioto 2000).

LATE ANTIQUE URBAN PATTERNS AND WATER SUPPLY

While it is possible to draw some general patterns for the evolution of the Roman cities in *Hispania*, the low number of examples for *Tingitania* makes this a more difficult task, and in fact it is easier to assess each case individually.

For some sites (*Banasa*, *Sala*, *Thamusida*), as mentioned, we have little or no evidence for urban continuity after the withdrawal of the frontier to the coastal strip. For a couple (*Zilil* and *Tamuda*) we have evidence of destruction followed by lack of urban activity at the time of the Vandal invasion. This leaves four case studies: Ceuta, *Lixus*, Tangiers, and *Volubilis*.

Ceuta evolved from a small settlement to a full urban nucleus during Late Antiquity, especially with the establishment of the Byzantine base. For Ceuta, water availability did not dictate the layout of the settlement, as other this was secondary to its importance as a military harbour within the greater Byzantine strategy for the West. In this sense, it may have not been a pure coincidence that the aqueduct led directly to the harbour area, where the *garum* factories were located. In the case of Tangiers, it would appear that after the fourth century the settlement clustered in the upper fortified area, under the current *alcazaba*, although there is no archaeological proof whatsoever. There might have been a Christian suburb, linked to a funerary basilica, and it might have been the seat of a bishop, but it seems that Tangiers did not have a set of walls (which might have favoured the development of Ceuta). During the Idrisid period, a new fortified settlement (*Tandja al-Baliya*) was built one kilometre away, although the core of the city shifted bach to its Roman location during the later Umayyad occupation (Siraj 1994a, 1994b). Water must have been secured from springs or cisterns, and not from the water table in both cases, and while in Ceuta the aqueduct had a primary role in the original selection of the site, in Tangiers it became (if it ever was) irrelevant.

Volubilis had an aqueduct which is generally assumed to have been out of use by the end of the fourth century. By this time the site had already began to shrink in size, and its core slowly shifted away from the plateau, abandoning the north-east quarter and the forum area, in favour of the slope of the hill towards the river and the water table, where it was possible to dig wells. This shift was further confirmed by the construction of a new wall across the old settlement (figure 89), leaving the area previously fed by the aqueduct outside and largely quarried. At the foot of the hill and during the Idrisid period a new set of elite buildings was established, including a set of baths, a reception room and a mosque, so the whole power balance of the settlement shifted towards it (Fentress and Limane 2010, Fentress 2017, Étienne 1960). With much more limited evidence, a similar development of shrinking of the fortifications and shift from the plateau towards the slope and the river is visible in *Lixus*, with a (late?) wall across the old site, and an active harbour suburb which continued active as a trading hub into the sixth century (Bernal Casasola 2008, 378, Mueden 2010, 116). In this case, however, the chronology of the aqueduct and of the shift cannot be correlated with certainty (Lenoir 1992).

STATE FORMATION AND WATER SUPPLY

Tingitania did not develop, as *Hispania* did, a unified polity that exercised its control over its territory in the immediate aftermath of the fall of Rome. In fact, it appears that the situation of power vacuum that ensued the withdrawal of the troops (and later the Vandal invasion) was the norm largely until the Islamic invasion. But it appears that there was division between the littoral cities and those of the highlands. The areas closer to the mountains seem to have been brought into the area of influence of the Berber kingdom of Masuna of Altava, which might have reached into *Volubilis*. Coastal towns like Tangiers and *Lixus* would have preserved closer links with the wider Mediterranean trade links, but without an evident centralising polity that could carry out large building projects (or, at least, of fortifications).[3] In any case, neither the highland kingdoms nor the coastal *civitates* appear to have acted in anyway on the old Roman aqueducts. In the Iberian Peninsula, the Justinianic conquest result-

[3] *CIL* VIII.9835: *Pro sal(ute) et incol(umitate) reg(is) Masunae gent(ium) Maur (orum) et Romanor(um) castrum edific(atum) a Masgivini pref(ecto) de Safar. Iidir proc(urator) castra Severian(a) quem Masuna Altava posuit, et Maxim(us) pr(ocurator) Alt(ava) prefec(it). P(ositum) p(rovinciae) CCCLXVIIII.*

ed in the establishment of a new province subordinate to Carthage, but none of the African cities other than Ceuta appear to have been under direct control, even if they were still within the Empire's sphere of commercial influence (Bernal Casasola 2008). In Ceuta we have seen already either a lack of interest in repairing or an untraceable maintaining of the aqueduct.

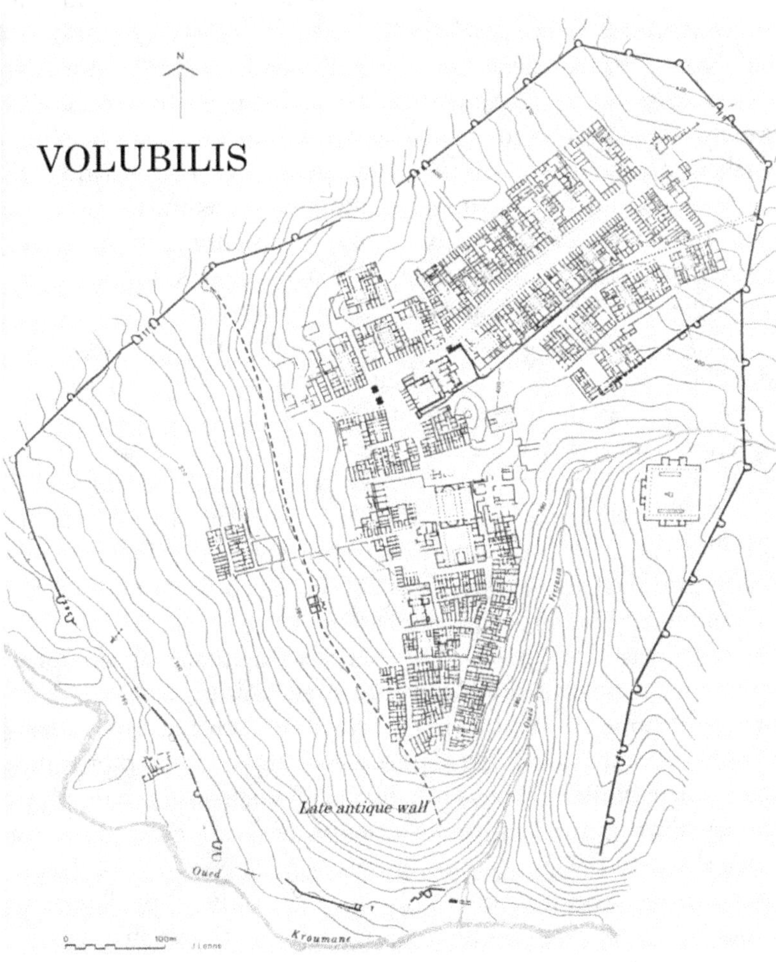

Figure 89: Plan of Volubilis (Euzennat 1989, fig. 134).

After the Islamic conquest, the Umayyad administration of *Tingitania* was focused around towns, despite the large semi-nomad Berber population beyond the coastal strip. Tangiers, Ceuta and *Volubilis* were important central places and active mints (Fentress 2017) from which to control the agricultural resources (Boone, et al. 1990). But beyond these facts, there is virtually nothing that can be said about the urbanism of *Tingitania* for the Umayyad period (Fenwick 2013), and it is likely that, much as it happened in Iberia until the late eighth century, the same urban patterns which had developed in Late Antiquity continued to develop.

From the moment of the conquest, around 700, up to the Berber revolt (740s) we find no significant investment by the state in any of these sites, nor any monumental attempts by the local elites. Furthermore, not even with the Idrisid period of state formation we see any investment in urban water supplies. In secondary sites like *Lixus* the evidence is limited; we know it was an important Idrisid centre as well (praised for its walls), but archaeology shows a gap between the fifth/sixth centuries and the eleventh (Brouquier-Reddé, El Khayari and Ichkhakh 2006). We only know of a building which could have been a basilica turned into a mosque, or an early mosque with later phases; the jury is still out (Aranegui Gascó 2001, 115). Here we can only assume that the cisterns continued to be the main source (El Khatib-Boujibar 1992). In *Volubilis* (renamed then Walili), which became the early capital when Idris I established an alliance with the local Berbers, there is the new elite complex which includes a hammam mentioned above, but it was not linked to any sort of wither long-distance or public water supply system.

The Idrisid's approach to the old Roman urban water supplies is the same as that of the Umayyads in al-Andalus, passive lack of interest, although in the Idrisid case we do not see a diversion or repair of old conduits for private purposes. This is, however, probably due to the lack of aqueducts in a reasonable state in the early ninth century. Even in their new foundations of the ninth century, Fez and Basra, there is a lack or organised public water supply systems. In the case of Fez, which substituted *Volubilis* as the Idrisid capital, the city is located near a river and in an area where the water table is easily reached and wells can be dug (Madani 2003). Conduits (*qanats*) and fountains would come later, in the eleventh century and introduced from al-Andalus by the Almoravids (Denoix 2008, 135-7).

CONCLUSIONS

Despite their geographical proximity and administrative links during the Roman period (and beyond), the degree of development of *Hispania* (as a whole) is not comparable to that of *Tingitania*. At a regional level it may be possible to make fair comparative studies between the north and south coasts of the straits of Gibraltar, but in terms of post-Roman urbanism and aqueducts there is no point of comparison. The underdeveloped urbanism, the prolonged period of political vacuum, the lack of centralising political attempts until the Islamic conquest and the overall volume of archaeological data clearly place both regions apart.

It is clear that aqueducts defined a Roman lifestyle in the province of *Tingitania* which did not continue into Late Antiquity. From the fourth century onwards we only have evidence for one functional aqueduct, that of Ceuta which was, furthermore, not fully developed as an urban centre until the Byzantine invasion. The one point of comparison between Iberia and Morocco is the lack of interest of the Islamic polities towards Roman public aqueducts or of water munificence in the Classical sense. In both cases it would only be much later in the Middle Ages when we see the development of integrated water supply systems. In terms of urbanism, *Lixus* and *Volubilis* show patterns of relocation away from areas which used to be supplied by aqueducts and heavy reliance on alternative systems such as wells, springs, or cisterns.

The increasing number of archaeological excavations and surveys currently being done in Morocco may, in the immediate future, cast more light on the evolution of the aqueducts and of the urbanism of its Roman sites. *Banasa* and *Thamusida* could still provide with more data about the course, origins and chronology of their aqueducts. Further research on *Sala* and *Lixus*, which show commercial activity into the Visigothic period, may lead to clearer views on the overall urbanism of these sites. The elephant in the room still is Tangiers the one city which could have the most interesting history and the one of which we know the least.

BIBLIOGRAPHY

ABBREVIATIONS AND PRIMARY SOURCES

Al-Makkari: Al-Makkari, *The History of the Mohammedan Dynasties in Spain*. Ed. and transl. Pascual de Gayangos, 1840. London: Oriental Translation Fund.

Arch.: Vitruvius, *De Architectura*. Ed. and transl. Frank Granger, 1933. Loeb Classical Library. Cambridge, MA: Harvard University Press.

Aq.: Frontinus, *De Aquis*. Ed. and transl. Jeffrey Henderson, 1925. Loeb Classical Library. Cambridge, MA: Harvard University Press.

Avienus: Avienus, *Ora Maritima*. Ed. Adolf Schulten, 1955. *Fontes Hispaniae Antiquae* I. Barcelona: University of Barcelona.

Bell. Goth.: Procopius, *De Bello Gothico*. Ed. and transl. Henry Dewing, 1914. Loeb Classical Library. Cambridge, MA: Harvard University Press.

Bicl.: John of Biclar, *Chronica*. Ed. Theodor Mommsen, 1894. *Monumenta Germaniae Historica, Auctores Antiquissimi* II. Berlin: Weidmann.

Chr. Caesaraug.: *Chronica Caesaraugustana*. Ed. Theodor Mommsen, 1894. *Monumenta Germaniae Historica, Auctores Antiquissimi* II. Berlin: Weidmann.

Chronica Rotensis. Ed. Gil Jiménez, 1985. *Crónicas asturianas*. Oviedo: University of Oviedo.

CIL: Corpus Inscriptionum Latinarum. Berlin: Prussian Academy of Sciences.

CIust: Codex Iustinianus. Ed. Paul Krueger, 1915. *Corpus Iuris Civilis,* v. 2. Berlin: Weidmann.

CNV: Vico Monteolivo, Jesús, Mari Cruz Cores Gomendio, and Gonzalo Cores Uria. 2006. *Corpus Nummorum Visigothorum ca. 575-714. Leovigildus – Achila.* Madrid.

Cont. Hisp.: *Continuatio Hispana.* Ed. Theodor Mommsen, 1894. *Monumenta Germaniae Historica, Auctores Antiquissimi* 11. Berlin: Weidmann.

CTheod: Codex Theodosianus. Ed. and transl. Clyde Pharr, 1952. *The Theodosian Code and Novels and the Sirmondian Constitutions.* Princeton: Princeton University Press.

Dig.: *Digestae Iustiniani.* Ed. Paul Krueger, 1915. *Corpus Iuris Civilis,* v. 1. Berlin: Weidmann.

DLH: Gregory of Tours, *Decem Libri Historiarum.* Ed. Bruno Krusch and Wilhelm Levison, 1951. *Monumenta Germaniae Historica Scriptores Rerum Merovingicarum* 1. Hannover: Hahnsche.

Etym.: Isidore of Seville, *Etymologiarum libri viginti sive origines.* Ed. Wallace Lindsay, 1911. Oxford: Clarendon.

Gregory, *Epist.*: Gregory the Great, *Epistulae.* Transl. John Martyn, 2004. *The Letters of Gregory the Great.* Medieval Sources in Translation 40. Toronto: Pontifical Institute of Mediaeval Studies.

Hist. Goth.: Isidore of Seville, *Historia de regibus Gothorum.* Ed. Theodor Mommsen, 1894. *Monumenta Germaniae Historica, Auctores Antiquissimi* 11. Berlin: Weidmann.

Hist. Lang.: Paul the Deacon, *Historia Langobardorum.* Ed. Ludwig Bethmann and Georg Weitz, 1878. *Monumenta Germaniae Historica Scriptores Rerum Langobardicarum et Italicarum.* Hannover: Hahnsche.

HWR: Julian of Toledo, *Historia Wambae Regis.* Ed. Jocelyn Hillgarth, 1976. *Corpus Christianorum: Series Latina,* 115. Turnhout: Brepols.

Hydat.: Hydatius, *Chronica subita.* Ed. and transl. Richard Burgess, 1993. *The Chronicle of Hydatius and the Consularia Constan-*

tinopolitana: Two Contemporary Accounts of the Final Years of the Roman Empire. Oxford: Oxford University Press.

Ibn Abd al-Hakam: *History of the Conquest of Spain.* Transl. John Harris Jones, 1858. *Ibn Abd el-Hakem, History of the Conquest of Spain*, Gottingen.

ICERV: Vives, José. 1942. *Inscripciones cristianas de la España romana y visigoda*. Barcelona: CSIC.

John Eph.: John of Ephesus: *Historia Ecclesiastica.* Ed. Ernest Brooks, 1935. *Corpus Christianorum Orientalium* 105. Leuven: Peeters.

Ktis.: Procopius, *Peri Ktismáton* (*Buildings*). Ed. and transl. Henry Dewig, 1914. Loeb Classical Library. Cambridge, MA: Harvard University Press.

Lib. Pont.: *Liber Pontificalis.* Transl. Raymond Davies, 2009. *The Book of the Pontiffs*. Liverpool Translated Texts for Historians. Liverpool: University of Liverpool.

LV: Lex Visigothorum. Ed. Karl Zeumer, 1902. *Monumenta Germaniae Historica Leges Nationum Germanicarum* 1. Hannover: Hahnsche.

Menand.: Menander Protector, *Historiae Fragmenta.* Transl. Roger Brockley, 1985, *The History of Menander the Guardsman*. Liverpool; Francis Cairns.

Nov.: *Novellae Iustiniani.* Ed. Paul Krueger, 1915. *Corpus Iuris Civilis*, v. 3. Berlin: Weidmann.

Paul. Nol., *Carm.*: Paulinus of Nola, *Carmina.* Online edition on The Latin Library, www.thelatinlibrary.com/paulinus.poemata.html (no source given)

Pliny, *Epist.*: Pliny the Younger, *Epistulae*. Ed. and transl. Betty Radice, 1969. Loeb Classical Library. Cambridge, MA: Harvard University Press.

Sid. Ap., *Epist.*: Sidonius Apollinaris, *Epistulae*. Ed. and transl. Williamm Anderson, 1936. Loeb Classical Library. Cambridge, MA: Harvard University Press.

Strabo: Strabo, *Geographica*. Ed. and transl. Horace Jones, 1924. Loeb Classical Library. Cambridge, MA: Harvard University Press.

Var.: Cassiodorus, *Epistulae Theodoricanae Variae*. Ed. Theodor Mommsen, 1894. *Monumenta Germaniae Historica Auctores Antiquissimi* 12. Berlin: Weidmann.

VHI: Astronomer, *Vita Hludowici Imperatoris*. Ed. Ernst Tremp, 1995. *Monumenta Germaniae Historica Scriptores Rerum Germanicarum* 64. Hannover: Hahnsche.

VSPE: *Vitae Sancotrum Patrum Emeritensium*. Transl. A. Fear, 1997. *Lives of the Visigothic Fathers*. Liverpool Translated Texts for Historians. Liverpool: Liverpool University Press.

Ximénez de Rada: Rodrigo Ximénez de Rada, *Historia de Rebus Hispaniae*. Ed. Juan Fernández Valverde and Juan Antonio Estévez Sola, 1999. *Corpus Christianorum Continuatio Mediaevalis* 72. Turnhout: Brepols.

REFERENCES

Abad Casal, Lorenzo, Sonia Guitiérrez Lloret, Blanca Gamo Parras, and Pablo Cánovas Guillén. 2008. "Una ciudad en el camino: pasado y futuro de El Tolmo de Minateda (Hellín, Albacete)." In *Recópolis y la ciudad en la época visigoda*. Zona Arqueológica 9, edited by Lauro Olmo Enciso, 323-36. Alcalá de Henares: Museo Arqueológico Regional.

Abadía Doñaque, José Carlos. 1995. "Algunos comentarios sobre el abastecimiento de agua a Caesar Augusta." *Cuadernos de Aragón* 23: 5-32.

Abel Vilela, Adolfo, and Felipe Arias Vilas. 1975. *Guía arqueológica romana de Lugo y su provincia*. Lugo: Diputación de Lugo.

Acero Pérez, Jesús. 2011. "Augusta Emerita." In *La gestión de los residuos urbanos en Hispania*. Anejos de Archivo Español de Arqueología 60, edited by Josep Anton Remolà Vallverdú and Jesús Acero Pérez, 157-80. Mérida: CSIC.

——. 2018. *La gestión de los residuos en Augusta Emerita. Siglos I a.C. - VII d.C.* Anejos de Archivo Español de Arqueología 82. Mérida: CSIC.

Acién Almansa, Manuel. 1987. "Madinat al-Zahra en el urbanismo musulmán." *Cuadernos de Madinat al-Zahra* 1: 11–26.

Adserías Sans, María, César Augusto Pociña López, and Josep Anton Remolà Valverdú. 2000. "L'habitat suburbà portuari de l'antiga Tàrraco. Excavacions al sector afectat per el PERI 2 (Jaume I – Tabacalera)." In *Tàrraco 99. Arqueologia d'una capital provincial romana*, edited by Joaquín Ruiz de Arbulo, 137-54. Tarragona: ICAC.

Alarcão, Jorge de, and Robert Étienne. 1977. *Fouilles de Conimbriga*. Vol. I. Paris.

Alba Calzado, Miguel. 2001. "Mérida entre la tardoantigüedad y el Islam: datos documentados en el Área Arqueológica de Morería." In *La islamización de la Extremadura romana*. Cuadernos Emeritenses 17, edited by Fernando Valdés and Agustín Velázquez, 265-308. Mérida: MNAR.

——. 2004. "Evolución y final de los espacios romanos emeritenses a la luz de los datos arqueológicos (pautas de transformación de la ciudad Tardoantigua y Altomedieval)." In *Augusta Emerita. Territorios, espacios, imágenes y gentes en Lusitania romana*, edited by Trinidad Nogales, 209-55. Mérida: MNAR.

——. 2005. "La vivienda en Emerita durante la antigüedad tardía: propuesta de un modelo para Hispania." In *VI Reunió d'arqueologia cristiana hispànica. Les ciutats tardoantigues d'Hispania: Cristianització i topografia*, edited by Josep María Gurt Esparraguera and Albert Ribera Lacomba, 121–50. Barcelona: Institut d'Estudis Catalans.

——. 2007. "Contribuciones al estudio de las infraestructuras hidráulicas de Augusta Emerita." In *El agua y las ciudades romanas*, edited by Julio Mangas and Martínez Santiago, 147-82. Madrid: Ediciones 2007.

——. 2018. "Secuencias en la transformación de Augusta Emerita (Siglos I-IX): tres concepciones distintas de ser ciudad." In *Entre civitas y madina. El mundo de las ciudades en la Península Ibérica y en el norte de África (siglos IV-IX)*. Collection de la Casa de Velázquez 167, edited by Sabine Panzram, 51–74. Madrid: Casa de Velázquez.

Alba Calzado, Miguel, and Pedro Mateos Cruz. 2008. "El paisaje urbano de Emerita en época visigoda." In *Recópolis y la ciudad en la época visigoda*. Zona Arqueológica 9, edited by Lauro

Olmo Enciso, 261–73. Alcalá de Henares: Museo Arqueológico Regional.

Albiach Delscals, Rosa, Israel Espí Pérez, and Albert Ribera Lacomba. 2009. "El agua sacra y su vinculación con el origen y el desarrollo urbano de una fundación romana. El santuario (¿asklepeion?) de Valentia (Hispania)." In *Santuarios, oppida y ciudades: arquitectura sacra en el origen y desarrollo urbano del Mediterráneo occidental.* Anejos de Archivo Español de Arqueología 45, edited by Pedro Mateos Cruz, Sebastián Celestino Pérez, Antonio Pizzo and Trinidad Tortosa Rocamora, 417-48. Mérida: CSIC.

Albiach Delscals, Rosa, Carles Ballester Martínez, and Miquel Rosselló Mesquida. 1999. *Memòria de les excavacions arqueològiques de la campanya de desmuntatge d'estructures en el solar de l'Almoina (València). 9ª campanya (Maig-Octubre 1998).* Excavation report, Valencia: SIAM.

Alejandre Sánchez, Francisco Javier. 2018. "Consideraciones sobre la datación de morteros de cal mediante 14C." *PH: Boletín del Instituto Andaluz del Patrimonio Histórico* 93: 122-40.

Alföldy, Geza. 1992. "Die Inschrift des Aquäduktes von Segovia. Ein Vorbericht." *Zeitschrift für Papyrologie und Epigraphik* 94: 231–48.

Algarra Pardo, Víctor Manuel, Consuelo Matamoros, and Asunción Viñes Pérez. 1993. *Informe de la intervención arqueológica realizada en la calle Avellanas 11–13 de Valencia.* Excavation report, Valencia: SIAM.

Almagro Gorbea, Antonio. 1987. "Planimetría de ciudades hispanomusulmanas." *Al-Qantara* 8: 421–8.

Almagro Gorbea, Martín, and Luis Caballero Zoreda. 1977. "Las excavaciones realizadas a lo largo del acueducto de Segovia." In *Segovia. Symposium de arqueología romana*, 33-42. Barcelona: University of Barcelona.

Álvarez Asorey, Rubén, María Covadonga Carreño Cascón, and Enrique González Fernández. 2003. *Aqva Vrbi. Historia do abastecemento de auga á cidade de Lugo (época romana-século XX).* Trabaxos de arqueoloxía 1. Lugo: Concello de Lugo.

Álvarez García, Nuria, and Albert Ribera Lacomba. 2003. *Informe arqueológico de la excavación de l'Almoina (Valencia).* Excavation report, Valencia: SIAM.

Álvarez García, Nuria, Carmina Ballester Martínez, Yolanda Carrión Marco, Elena Grau Almero, Guillermo Pascual Berlanga, Guillem Pérez Jordà, Albert Ribera Lacomba, and Carmen Rodríguez Santana. 2005. "L'area productiva d'un edifici del fórum de Valentia al baix imperi." In *VI Reunció d'arqueologia cristiana hispànica. Les ciutats tardoantigues d'Hispania: Cristianització i topografía*, edited by Josep Maria Gurt Esparraguera and Albert Ribera Lacomba, 251–60. Barcelona: Institut d'Estudis Catalans.

Álvarez Martínez, José María. 1977. "En torno al acueducto de Los Milagros (Mérida)." In *Segovia. symposium de arqueología romana*, 49-60. Barcelona: University of Barcelona.

—. 2007. "El agua de Augusta Emerita." In *El agua y las ciudades romanas*, edited by Julio Mangas and Santiago Martínez, 183-212. Madrid: Ediciones 2007.

Ammar, Hakim. 2008. "À propos du nymphée de Sala." *L'Africa Romana* 17.1: 559-70.

Aquilué, Xavier. 1993. *La sede del Col·legi d'arquitectes. Una intervención arqueológica en el centro histórico de Tarragona.* Tarragona: Colegio de Arquitectos de Cataluña.

Aranda Gutiérrez, Fernando, Jesús Carrobles Santos, and José Luis Isabel Sánchez. 1997. *El sistema hidráulico romano de abastecimiento a Toledo.* toledo: Instituto de Estudios Toledanos.

Aranda Gutiérrez, Fernando, José Luis Sánchez Carcaboso, Esperanza Andrés Díaz, Francisco Rodríguez Martín, María Eugenia Polo García, Rafael Sánchez Crespo, and José Antonio Gutiérrez Gallego. 2006. "Las presas de abastecimiento en el marco de la ingeniería romana. Los casos de Proserpina y Cornalbo." *Mérida, excavaciones arqueológicas* 9: 471–536.

Aranda Gutiérrez, Fernando, María Eugenia Polo García, José Luis Sánchez Carcaboso, and Esperanza Andrés Díaz. 2017. "El abastecimiento de agua a las ciudades en Hispania, de época romana a la medieval." In *Wasserversorgung in Toledo un Wissenvermittlung von der Antike ins Mittelalter.* Iberia Archaeologica 19, edited by Thomas Schattner and Fernando Valdés, 185-202. Berlin: DAI.

Aranegui Gascó, Carmen. 2001. *Lixus. Colonia fenicia y ciudad púnico-mauritana. Anotaciones sobre su ocupación medieval.* Valencia: University of Valencia.

Arce Martínez, Javier. 2008. "La inscripción del puente de Mérida de época del rey Eurico (483 d.C.)." *Pyrenae* 39.2: 121–26.

—. 2010. "Recópolis, la ciudad fantasma." In *Doctrina a magistro discipulis tradita. Estudios en homenaje al profesor Dr. Luis García Iglesias,* edited by Adolfo Domínguez and Gloria Mora Rodríguez, 373-94. Madrid: Autonomous University.

Arce Sainz, Fernando. 2015. "La supuesta basílica de San Vicente en Córdoba: de mito histórico a obsesión historiográfica." *Al-Qantara* 36.1: 11–44.

Arenillas Parra, Miguel, and Marisa Barahona Oviedo. 2008. "Una revisión general del suministro de agua a Toledo en época romana." In *IV Congreso de las Obras Públicas en la Ciudad Romana,* 211–34. Madrid: Colegio de Ingenieros Técnicos de Obras Públicas.

Argente Oliver, José Luis, and Adelia Díaz Díaz. 1980. "Campaña 1977-78: el acueducto romano." In *Tiermes I.* Excavaciones Arqueológicas de España vol. 111, edited by José Luis Argente Oliver, 193-236. Madrid: Ministry of Culture.

Arharbi, Rachid, and Eliane Lenoir. 2006. "Recherches sur le quartier méridional de Banasa." *L'Africa Romana* 16.4: 2141–56.

Arias Vilas, Felipe, and Antonio de Vega Rodríguez. 1997. "Las termas Romanas de Lugo." In *Termalismo Antiguo,* edited by María Jesús Pérex Agorreta, 345-52. Madrid: UNED.

Arjona Castro, Antonio. 2001. *Córdoba en la historia de al-Ándalus. Desarrollo, apogeo y ruina de la Córdoba Omeya.* Córdoba: Royal Academy of Córdoba.

Bagust, J., J. Fletcher, and G. Morgan. 2011. "Mortars and plasters." In *Leptiminus (Lamta) Report no. 3. The Field Survey.* JRA

Supplement Series 87, edited by David Stone, David Mattingly and Nejib Ben Lazreg, 454-62. Portsmouth: JRA.

Beaujard, Briguitte. 1996. "L'évêque dans la cité en Gaul aux Ve et VIIe siècles." In *La fin de la cité antique et le début de la cité médiévale*, edited by Claude Lepelly, 127-46. Bari: Edipuglia.

Bedon, Robert. 1997. *Les aqueducs de la Gaulle romaine et des les regions voisines.* Caesarodunum 31. Limoges: Presses Universitaires de Limoges.

Beltrán de Heredia, Julia. 2002a. "Fullonica and tinctoria. Fabrics, dyes and laundry in the ancient Roman colony." In *From Barcino to Barchinona (1st to 7th centuries). The Archaeological Remains of the Plaça del Rei in Barcelona*, edited by Julia Beltrán de Heredia, 48-57. Barcelona: Museo de Historia de la Ciudad.

—. 2002b. "A garum and salt fish factory in Barcino." In *From Barcino to Barchinona (1st to 7th centuries). The Archaeological Remains of the Plaça del Rei in Barcelona*, edited by Julia Beltrán de Heredia, 58-65. Barcelona: Museo de Historia de la Ciudad.

—. 2002c. "Continuity and change in the urban topography. Archaeological evidence of the north-east quadrant of the city." In *From Barcino to Barchinona (1st to 7th centuries). The Archaeological Remains of the Plaça del Rei in Barcelona*, edited by Julia Beltrán de Heredia, 96-111. Barcelona: Museo de Historia de la Ciudad.

—. 2008. "Barcino durante la antigüedad tardía." In *Recópolis y la ciudad en la época visigoda.* Zona Arqueológica 9, edited by Lauro Olmo Enciso, 274-291. Alcalá de Henares: Museo Arqueológico Regional.

Beltrán de Heredia, Julia, and César Carreras Monfort. 2011. "Barcino." In *La gestión de los residuos urbanos en Hispania.* Anejos de Archivo Español de Arqueología 60, edited by Josep Anton Remolà Vallverdú and Jesús Acero Pérez, 233-54. Mérida: CSIC.

Beltrán Llorís, Miguel. 2007. "Introducción histórica." In *Zaragoza. Colonia Caesar Augusta*, edited by Francisco Beltrán Llorís, 3-14. Rome: L'Erma di Bretschneider.

Benoit, Paul, and Josephine Rouillard. 2000. "Medieval hydraulics in France." In *Working with Water in Medieval Europe. Technology and Resource-use*, edited by Paolo Squatriti, 161–216. Leiden: Brill.

Bernal Casasola, Darío. 1996. "Le anfore tardo-romane attestate a Ceuta (Septem Fratres, Mauretania Tingitana)." *L'Africa Romana* 11.3: 1191–1233.

—. 2008. "Las ciudades del Fretum Gaditanum tardoantiguo. Pesquerías y comercio transmediterráneo en época bizantina." In *Recópolis y la ciudad en la época visigoda.* Zona Arqueológica 9, edited by Lauro Olmo Enciso, 362-83. Alcalá de Henares: Museo Arqueológico Regional.

—. 2018. "Continuidad y cesura en las ciudades tardorromanas del Estrecho de Gibraltar." In *Entre civitas y madina. El mundo de las ciudades en la Península Ibérica y en el norte de África (ss. IV-IX).* Colección de la Casa de Velázquez 167, edited by Sabine Panzram and Lauent Callegarin, 105-18. Madrid: Casa de Velázquez.

Bernal Casasola, Darío, and José Manuel Pérez Rivera. 2000. "La ocupación bizantina de Septem. Análisis del registro arqueológico y propuestas de interpretación." In *V Reunió d'arqueologia cristiana hispànica*, edited by Josep Maria Gurt Esparraguera and Núria Tena, 121–34. Barcelona: Institut d'Estudis Catalans.

Bernal Casasola, Darío, José Manuel Pérez Rivera, Santos Carvajal Montero, José Angel Expósito Alvarez, and Lourdes Lorenzo Martínez. 2005. "El urbanismo de "Septem" en la antigüedad tardía: novedades de las actuaciones arqueológicas en el paseo de las Palmeras." In *VI Reunió d'arqueologia cristiana hispànica: les ciutats tardoantigues d'Hispania: cristianització i topografia*, edited by Josep Maria Gurt Esparraguera and Albert Ribera Lacomba, 435-448. Barcelona: Institut d'Estudis Catalans.

Bernal Casasola, Darío, Baraka Raissouni, Macarena Bustamante, Antonio Sáez, José Juan Díaz, Joé Lagóstena, and Macarena

Lara. 2012. "La datación de Tamuda. Asentamiento púnico, ciudad mauritana y castellum romano: novedades estratigráficas." *L'Africa Romana* 19.3: 2443-78.

Bestué Cardiel, Isabel, and Ignacio González Tascón. 2006. "Roman siphons in Spain." In *Cura Aquarum in Ephesus*, edited by Gemma Wiplinger, 311–8. Leuven: Peeters.

Blackwell, Bonnie. 2006. "Electron Spin Resonance (ESR) dating in karst environments." *Acta Carsologica* 35.2-3: 123-53.

Blasco, Concepción, Raúl Durán, Asunción Millán, Tomás Calderón, Pedro Benéitez, José Arribas, and Manuel Bendala. 1993. "Datación por termoluminiscencia de la arquitectura de ladrillo: el caso de Mérida, resultados y problemática." *Cuadernos de prehistoria y arqueología* 20: 239-54.

Blyth, P. H. 1999. "The consumption and cost of fuel in hypocaust baths." In *Roman Baths and Bathing*. JRA supplement series number 37, edited by Janet DeLaine and David Johnson, 87-93. Portsmouth: JRA.

Bonilla Santander, Óscar, and Begoña Serrano Arnáez. 2012. "El paisaje rural romano en torno al Municipium Turiaso (España)." In *Arqueologia de un paisaje en transición. Antigüedad Tardía y Alta Edad Media*, edited by Pilar Diarte Blasco and Alejandro Martín López, 95-100. Zaragoza: University of Zaragoza.

Bonnet, Charles, and Julia Beltrán de Heredia. 2002. "The origins and evolution of the episcopal buildings in Barcino: from the early Christian period to the Visigothic era." In *From Barcino to Barchinona (1st to 7th centuries). The Archaeological Remains of the Plaça del Rei in Barcelona*, edited by Julia Beltrán de Heredia, 74-95. Barcelona: Museo de Historia de la Ciudad.

Boone, James, Emlen Myers, Redman, and Charles. 1990. "Archeological and historical approaches to complex societies: the Islamic states of Medieval Morocco." *American Anthropologist* 92.3: 630-46.

Bosch Puche, Francisco, Josep Maria Macias Solé, Joan Menchon Bes, Andreu Muñoz Melgar, and Immaculada Teixell Navarro. 2005. "La transformació urbanística de l'acròpolis de Tarracona:

avanç de les excavacions del Pla director de la Catedral de Tarragona (2000-2002)." In *VI Reunió d'arqueologia cristiana hispànica. Les ciutats tardoantigues d'Hispania: cristianització i topografia*, edited by Josep Maria Gurt Esparraguera and Albert Ribera Lacomba, 167-74. Barcelona: Institut d'Estudis Catalans.

Bouet, Alain. 2003. *Les thermes privés et publics en Gaul Narbonnaise.* Rome: École Française de Rome.

Brouquier-Reddé, Véronique, Abdelaziz El Khayari, and Abdelfattah Ichkhakh. 2006. "Lixus, de l'époque phénicienne à la période médiévale: le quartier dit des temples." *L'Africa Romana* 16.4: 2157-74.

Brunn, Christer. 2002a. "Water legislation in the ancient world (c. 2200 BC – c. AD 500)." In *Handbook of Ancient Water Technology*, edited by Örjan Wikander, 539-606. Leiden: Brill.

—. 2002b. "Water shortage and surplus in the ancient world." In *Cura Aquarum in Sicilia*, edited by Gemma Jansen, 215-34. Leuven: Peteers.

—. 2012. "Roman emperors and legislation on public water in the Roman Empire: clarifications and problems." *Water History* 4.1: 11–34.

Buchberger, Erica. 2013. *From Romans to Goths and Franks: Ethnic Identities in Sixth- and Seventh-Century Spain and Gaul.* DPhil thesis, Oxford: Oxford University.

Burès, Lurdes, Montse García, and Josep Maria Macias Solé. 1998. "Un aqüeducte soterrani a Tàrraco." *Empùries* 51: 183-96.

Burgalhão, Jacinta. 2009. "Lisboa Islâmica: uma realidade em construção." *Xelb* 9: 379-93.

Caballos Rufino, Antonio, José Manuel Rodríguez Hidalgo, and Juan Marín Fatuarte. 1999. *Itálica arquelógica.* Seville: University of Seville.

Cabrera Tejedor, Carlos. 2016. *From Hispalis to Ishbiliyya: the ancient port of Seville, from the Roman Empire to the end of the Islamic period (45 BC - AD 1248).* DPhil thesis, Oxford: Oxford University.

Campomanes Alvaredo, Emilio. 2006. "El acueducto de la Legio VII Gemina y sus producciones latericias." In *Arqueología militar romana en España II: producción y abastecimiento en el ámbito militar*, edited by Ángel Morillo Cerdán, 439-54. León: University of León.

Canto, Alicia. 1978. "El acueducto romano de Itálica." *Madrider Mitteilungen* 20: 282-338.

—. 2001. "Fuentes árabes para la Mérida Romana." In *La islamización de la Extremadura romana.* Cuadernos Emeritenses 17, edited by Fernando Valdés and Agustín Velázquez, 9-86. Mérida: MNAR.

Carmona Berenguer, Silvia, Maudilio Moreno Almenara, and Marina González Viseda. 2008. "Un basurero califal con piezas de vidrio procedente del acueducto de la estación de autobuses de Córdoba." *Anejos de Anales de Arqueología Cordobesa* 1: 233-46.

Carver, Martin. 1996. "Transitions to Islam: Urban rôles in the east and south Mediterranean, fifth to tenth centuries AD." In *Towns in Transition: Urban Evolution in Late Antiquity and the Early Middle Ages*, edited by Neil Christie and Simon Loseby, 184-212. Aldershot: Scholar Press.

Casabona Sebastián, Juan Francisco. 2014. "Los orígenes romanos de la catedral de Tarazona (España)." In *XVIII CIAC: Centro y periferia en el mundo clásico*, edited by José María Álvarez, Trinidad Nogales and Isabel Rodà, 1861–4. Mérida: MNAR.

Casabona Sebastián, Juan Francisco, and Javier Delgado Ceamanos. 1991. "Informe de la excavación del solar c/ M. Carrillo, c/ Universidad y c/ Órgano (Zaragoza)." In *Arqueología Aragonesa 1988-1989*, 337-9. Zaragoza: Departamento de Cultura y Educación.

Casagrande, Massimo. 2011. "Leggi prediali e regolamenti cittadini: realizzazione e gestione degli acquedotti africani." In *Aquam perducendam curavit. Captación, uso y administración del agua en las ciudades de la Bética y el occidente romano*, edited by Lázaro Lagóstena Barrios, José Luis Cañizar Palacios and Lluís Pons Pujol, 461–72. Cádiz: University of Cádiz.

Casal García, María Teresa. 2008. "Características generales del urbanismo cordobés de la primera etapa emiral: el arrabal de Saqunda." *Anales de Arqueología Cordobesa* 1: 109-34.

Casimiro, Sílvia, and Rodrigo Banha da Silva. 2013. "Enterramentos infantis tardoantigos na rua de San Nicolau (Lisboa)." In *Arqueologia em Portugal. 150 anos*, edited by José Morais Arnaud, Andrea Martins and César Neves, 859-63. Lisbon: Associação de Arqueólogos Portugueses.

Castro García, Mari Mar. 2017. "Modelos de abastecimiento urbano de aguas en la Bética romana: las cisternas." *Espacio, Tiempo y Forma, Serie II* 30: 97-124.

Castro Orellana, Jonathan, and Joan Roig Ribas. 2009. "Nuevas evidencias sobre el abastecimiento de agua a la ciudad romana de Ebusus. El conjunto hiráulico del Camí de Cas Ferró." *Fites* 9: 19-31.

Castro Priego, Manuel, and Amaya Gómez de la Torre-Verdejo. 2008. "La actividad artesanal en Recópolis: la producción de vidrio." In *Recópolis y la ciudad en la época visigoda*. Zona Arqueológica 9, edited by Lauro Olmo Enciso, 117-28. Alcalá de Henares: Museo Arqueológico Regional.

Chalmeta Gendrón, Pedro. 1994. *Invasión e islamización. La sumisión de Hispania y la formación de al-Andalus*. Colección al-Andalus 10. Madrid: Mapfre.

Chalon, Michel, Georges Devallet, Paul Force, Michel Griffe, Jean-Marie Lassère, and Jean-Noël Michaud. 1985. "Memorabile factum. Une célébration de l'évergétisme des rois vandales dans l'Anthologie latine." *Antiquités africaines* 21: 207-262.

Civera Gómez, Manuel. 2004. "Aproximació a l'evolució urbana del Sagunt antic." *Arse* 38: 81–105.

—. 2008. "Els aqüeductes de la ciutat de Saguntum." *Arse* 42: 193-225.

Clapés Salmoral, Rafael. 2013. "Un baño privado en el arrabal occidental de Madinat Qurtuba." *Arqueología y Territorio Medieval* 20: 97-128.

Clarke, Nicola. 2012. *The Muslim Conquest of Iberia. Medieval Arabic Narratives*. London: Routledge.

Coates-Stephens, Robert. 1998. "The walls and aqueducts of Rome in the Early Middle Ages, AD 500-1000." *Journal of Roman Studies* 88: 166-89.

—. 2003a. "Gli acquedotti in epoca tardoantica nel suburbio." In *Suburbium. Il suburbia di Roma dalla crisi del sistema delle ville a Gregorio Magno*, edited by Philippe Pergola, Riccardo Santangeli Valenzani and Rita Volpe, 415-36. Rome: École Française de Rome.

—. 2003b. "The water-supply of early medieval Rome." In *Technology, Ideology, Water: From Frontinus to the Renaissance and Beyond*, edited by Christer Brunn and Ari Saastamoinen, 81–114. Rome: Finnish Institute at Rome.

—. 2004. *Porta Maggiore: Monument and Landscape. Archaeology and Topography of the Southern Esquiline from the Late Republican period to the Present.* Rome: British School at Rome.

Cordero Ruiz, Tomás, and Isaac Sastre de Diego. 2010. "El yacimiento de Casa Herrera en el contexto del territorio emeritense (siglos IV-VIII)." In *Espacios urbanos en el occidente mediterráneo, ss. VI-VIII*, edited by Alfonso García, 211–8. Toledo: Toletum Visigodo.

Correia, Virgílio Hipólito. 2010. "O forum de Conimbriga e a evoluçao do centro urbano." In *Ciudad y foro en Lusitania Romana*, edited by Trinidad Nogales Basarrate, 89-106. Badajoz.

Correia, Virgílio Hipólito, Adriaan De Man, and Maria Pilar Reis. 2011. "A Propósito De Uma Obra Recente Sobre o Período Tardo-Antigo e Medieval Em Conímbriga." *Conimbriga: Revista De Arqueologia* 50: 127–46.

Correia, Virgílio Hipólito, and Maria Pilar Reis. 2000. "As termas de Conimbriga: tipologías arquitectónicas e integração urbana." In *Termas romanas en el occidente del imperio. II Coloquio internacional de arqueología en Gijón*, edited by Carmen Fernández Ochoa and Virginia García-Entero, 271–80. Gijón: Ayuntamiento de Gijón.

Cortés, Rodolfo, Cristina Benet, and Alejandro Bermúdez. 1989. *Sobre los acueductos de Tarraco.* Vol. 1, in *XIX Congreso Nacional de Arqueología*, 1091–1100. Zaragoza.

Cressier, Patrice. 2001. “El acarreo de obras antiguas en la arquitectura islámica de primera época.” In *La islamización de la Extremadura romana.* Cuadernos Emeritenses 17, edited by Fernando Valdés and Agustín Velázquez, 309-34. Mérida: MNAR.

Creswell, Keppel Archibald. 1940. *Early Islamic Architecture. Umayyads, Early Abbasids and Tulunids. Part two: Early Abbasids, Umayyads of Cordova, Aghlabids, Tulunids and Samanids AD 751–905.* Oxford: Clarendon.

Crone, Patricia. 2005 [2004]. *Medieval Islamic Political Thought.* 2nd. Edinburgh: Edinburgh University Press.

Crow, Jim, Jonathan Bardill, and Richard Bayliss. 2008. *The Water Supply of Byzantine Constantinople.* JRS Monograph 11. London: The Roman Society.

Cuomo, Serafina. 2007. *Technology and Culture in Greek and Roman Antiquity.* Cambridge: Cambridge University Press.

Curchin, Leonard. 2014. “The role of civic leaders in late antique Hispania.” *Studia Historica. Historia Antigua* 32: 281–304.

De Man, Adriaan. 2007a. “A muralha tardia de Conimbriga.” In *Murallas de ciudades romanas en el occidente del imperio. Lucus Augusti como paradigma*, edited by Antonio Rodríguez Colmenero and Isabel Rodá de Llanza, 699-714. Lugo: Diputación de Lugo.

—. 2007b. “The late urban fort of the Bico da Muralha, Conimbriga (Portugal).” *Mediterranean Archaeology and Archaeometry* 7.1: 3-14.

—. 2011. *Defesas Urbanas Tardias da Lusitânia.* Stvdia Lusitania 6. Mérida: MNAR.

de Sepúlveda, Eurico, Ana Vale, Vítor Sousa, Natalina Guerreiro, and Vitor Santos. 2002. “A cronologia do circo de Olisipo: a Terra Sigillata.” *Revista portuguesa de arqueologia* 5.2: 245-275.

DeLaine, Janet. 1997. *The Baths of Caracalla: a Study in the Design, Construction and Economics of Large-Scale Building Projects in Imperial Rome.* JRA supplement series 25. Portsmouth: JRA.

Delgado Molina, Pedro Ángel. 2006. "Excavación de un área cristiana en época andalusí. Intervención arqueológica llevada a cabo en la calle B. Arias Montano c.v. a Avda. Reina Sofía (Mérida)." *Mérida: Excavaciones arqueológicas* 9: 285-312.

Deniaux, Élizabeth. 2011. "L'aqueduc de Dyrrachium, construction et restauration." In *Le réseaux d'eau courante dans l'antiquité. Réparations, modifications, réutilisations, abandon, récupération*, edited by Catherine Abadie-Reynal, Samuel Provost and Pascal Vipard. Rennes: Presses universitaires de Rennes.

Denoix, Sylvie. 2008. "Founded cities of the Arab world from the seventh to the eleventh centuries." In *The City in the Islamic World*, edited by Salma Jayyusi, 115-41. Leiden: Brill.

Dessales, Henri. 2011. "Entretien et restauration des aqueducs à Rome, au regard du traité du Frontin." In *Le réseaux d'eau courante dans l'antiquité. Réparations, modifications, réutilisations, abandon, récupération*, edited by Catherine Abadie-Reynal, Samuel Provost and Pascal Vipard, 13-26. Rennes: Presses universitaires de Rennes.

Detry, Cleia, João Cardoso, and Virgílio Hipólito Correia. 2014. "What did the Romans and Moslems eat in Conimbriga (Portugal)? The animal bones from the 1990's excavations." In *Proceedings of the First Zooarchaeology Conference in Portugal, At Faculty of Letters, University of Lisbon.* BAR IS 2662, edited by Cleia Detry and Rita Dias, 97-110. Oxford: Archaeopress.

Diarte Blasco, Pilar. 2011. *Morfología del cambio urbano en Hispania: La adaptación de los espacios públicos romanos a la nueva realidad tardoantigua.* PhD Thesis, Zaragoza: University of Zaragoza.

Díaz Díaz, Adelia, and Inmaculada Argente Oliver. 1984. "Edificio público número 19: castellum aquae." In *Tiermes II (Campañas de 1979 y 1980).* Excavaciones Arqueológicas de España 128, edited by Inmaculada Argente Oliver, Carlos Casa Martínez, Adelia Díaz Díaz, Victor Férnandez Martínez, Alfonso

González Uceda and Elías Terés Navarro, 15-52. Madrid: Ministerio de Cultura.

Díaz García, Moises, Montserrat García Noguera, and Josep Maria Macias Solé. 2000. "Les termes publiques de Tarragona: excavacions en el carrer de Sant Miquel, núm. 33." In *Tàrraco 99. Arqueologia d'una capital provincial romana*, edited by Joaquín Ruíz de Arbulo, 111–33. Tarragona: ICAC.

Díaz Martínez, Pablo. 2011. *El Reino Suevo (411–585)*. Madrid: Akal.

DiSegni, Leah. 2002. "The water supply of Roman and Byzantine Palestine in literary and epigraphical sources." In *The Aqueducts of Israel.* JRA Supplement Series 46, edited by David Amit, Joseph Patrich and Yizhar Hirschfeld, 37-67. Portsmouth: JRA.

Duncan-Jones, Richard. 1985. "Who paid for public buildings in Roman cities." In *Roman Urban Topography in Britain and the Western Empire*, edited by Francis Grew and Brian Hobley, 28-33. London: Council for British Archaeology.

Duval, Mathieu, Christophe Falguères, Rainer Grün, Qingfeng Shao, Maxime Aubert, Dolo, Jean-Michel Jean-Jacques Bahain, Jordi Agustí, Bienvenido Martínez-Navarro, Paul Palmqvist, and Isidro Toro-Moyano. 2012. "On the limits of using combined U-series/ESR method to date fossil teeth from two Early Pleistocene archaeological sites of the Orce area (Guadix-Baza basin, Spain)." *Quaternary Research* 77.3: 482-91.

Duval, Mathieu, Rainer Grün, Christophe Falguères, Jean-Jacques Bahain, and Jean-Michel Dolo. 2009. "ESR dating of Lower Pleistocene fossil teeth: Limits of the single saturating exponential (SSE) function for the equivalent dose determination." *Radiation Measurements* 44: 477–482.

Egea Vivancos, Alejandro. 2002. "Características principales del sistema de captación, abastecimiento, distribución y evacuación de aguas de Carthago Nova." *Empuries* 53: 13-28.

Egea Vivancos, Alejandro, Enrique Ruiz Valderas, and Jaime Vizcaíno Sánchez. 2011. "Carthago Nova." In *La gestión de los residuos urbanos en Hispania.* Anejos de Archivo Español de

Arqueología 60, edited by Josep Anton Remolà Vallverdú and Jesús Acero Pérez, 281–9. Mérida: CSIC.

El Khatib-Boujibar, Naïma. 1992. "Le problème de l'alimentation en eau à Lixus." In *Lixus: Actes du colloque de Larache (8-11 novembre 1989).* Collection de l'École Française de Rome 166, 305-23. Rome: École Française de Rome.

Elbl, Martin. 2013. *Portuguese Tangier (1471–1662). Colonial Urban Fabric as a Cross-Cultural Skeleton.* Toronto: Baywolf Press.

Elices Ocón, Jorge. 2017. *El pasado pre-islámico en al-Andalus: fuentes árabes, recepción de la antigüedad y legitimación en época omeya (ss. VIII-X).* PhD thesis, Madrid: Autonomous University.

Escorza, Carlos. 1997. "El agua de la fuente de San Juan de Baños (Palencia): Carácter y variación estacional." In *Termalismo antiguo*, edited by María Jesús Pérex Agorreta, 67-76. Madrid: UNED.

Escudero Escudero, Francisco, and María Pilar Galve Izquierdo. 2011. "Caesaraugusta." In *La gestión de los residuos urbanos en Hispania.* Anejos de Archivo Español de Arqueología 60, edited by Josep Anton Remolà Vallverdú and Jesús Acero Pérez, 255-80. Mérida: CSIC.

Esmonde Cleary, Simon. 2013. *The Roman West, AD 200-500. An Archaeological Study.* Cambridge: Cambridge University Press.

Étienne, Robert. 1960. *Le quartier nord-est du Volubilis.* Paris: Baccard.

Étienne, Robert, and Jorge de Alarcão. 1974. *O aqueduto romano de Conímbriga.* Vol. 1, in *Actas do III Congreso Nacional de Arqueologia*, 283-92. Oporto.

Euzennat, Maurice. 1989. *Le limes de Tingitane. La frontière méridionale.* Paris: CNRS.

Evan, Harry. 1994. *Water Distribution in Ancient Rome. The Evidence of Frontinus.* Ann Harbour: University of Michican Press.

Fabre, Guilhem, Jean-Luc Fiches, and Jean-Luc Paillet. 1991a. *L'aqueduct de Nîmes et le Pont du Gard.* Nîmes: CNRS.

—. 1991b. "The aqueduct of Nîmes and the Pont du Gard." *Journal of Roman Archaeology* 4: 63-88.

Fahlbusch, Henning. 1991. "Maintenance problems in ancient aqueducts." In *Future Currents in Aqueduct Studies*, edited by Trevor Hodge, 7-14. Leeds: Francis Cairns.

Faro Carballa, José Antonio, María García-Barberena Unzu, and Mercedes Unzu Urmeneta. 2008. "Pamplona y el Islam: nuevos testimonios arqueológicos." *Trabajos de Arqueología de Navarra* 20: 229-284.

Feijoo Martínez, Santiago. 2005. "Las presas y los acueductos de agua potable. Una asociación incompatible en la antigüedad. El abastecimiento a Augusta Emerita." In *Augusta Emerita. Territorios, espacios, imágenes y gentes en Lusitania Romana*, edited by Trinidad Nogales Basarrate, 171–205. Mérida: MNAR.

—. 2006. "Las presas y el agua potable en época romana: dudas y certezas." In *Nuevos elementos de ingeniería romana. III congreso de las obras públicas romanas*, edited by Isaac Moreno Gallo, 145-60. Madrid: Junta de Castilla y León.

Feijoo Martínez, Santiago, and Diego Gaspar Rodríguez. 2019. "Primeros avances sobre el estudio de dos nuevos tramos de conducciones en Mérida." *Cimbra* 415: 24-33.

Fentress, Elizabeth. 2017. "Idris I and the Berbers." In *The Aghlabids and their Neighbours*, edited by Glaire Anderson, Corisande Fenwick and Mariam Rosser-Owen, 514-30. Leiden: Brill.

Fentress, Elizabeth, and Hassan Limane. 2010. "Excavations in medieval settlements at Volubilis. 2000-2004." *Cuadernos de Madinat al-Zahra* 7: 105-22.

Fenwick, Corisande. 2013. "From Africa to Ifriqiya: settlement and society in early Medieval North Africa (650–800)." *Al-Masaq* 25.1: 9–33.

Fernandes, Lídia. 2013. "Teatro Romano de Olisipo: a marca do novo poder romano." In *Arqueologia em Portugal. 150 anos*, edited by José Morais Arnaud, Andrea Martins and César Neves, 765-73. Lisbon: Associação de Arqueólogos Portugueses.

Fernández Casado, Carlos. 1949. "La conducción romana de aguas a Almuñécar." *Archivo Español de Arqueología* 77: 313-33.

—. 2008 [1972]. *Acueductos romanos en España.* 2nd. Madrid: Colegio de Ingenieros de Caminos.

Fernández Colmenero, Alicia, and Celso Rodríguez Cao. 2012. "Anastilosis virtual de A domus do Mitreo do Lvcvs Avgvsti." *Virtual Archaeology Review* 3.5: 104-08.

Fernández Delgado, Aitor, Javier Martínez Jiménez, and Carlos Tejerizo García. 2013. "Old and new elites in the Visigothic kingdom (550-650 AD)." In *Tough Times: The Archaeology of Crisis and Recovery. Proceedings of the GAO Annual Conferences 2010 and 2011.* BAR IS 2478, edited by Elsbeth Van der Wilt and Javier Martínez Jiménez, 161–70. Oxford: Archaeporess.

Fernández Fernández, Adolfo. 2014. *El comercio tardoantiguo (ss. IV-VII) en el Noroeste peninsular a través del registro cerámico de la ría de Vigo.* Roman and Late Antique Mediterranean Pottery 5. Oxford: Archaeopress.

Fernández Martínez, Concepción, and Joan Gómez Pallarés. 2001. "Hermenegildo, ¿para siempre en Sevilla? Una nueva interpretación de IHC n. 76 = ILCV n.50." *Gerión* 19: 629-58.

Fernández Ochoa, Carmen, Ángel Morillo Cerdán, and Jorge López Quiroga. 2005. "La dinámica urbana en las ciudades de la fachada noratlántica y del cuadrante noroeste de Hispania durante el Bajo Imperio y la Antigüedad Tardía (siglos III-VII d.C.)." In *VI Reunió d'arqueologia cristiana hispànica. Les ciutats tardoantigues d'Hispania: cristianització i topografía,* edited by Josep Maria Gurt Esparraguera and Albert Ribera Lacomba, 91–119. Barcelona: Institut d'Estudis Catalans.

Fernández, Damián. 2006. "What is the De fisco Barcinonensi about?" *Antiquité Tardive* 14: 217-24.

—. 2017. *Aristocrats and Statehood in Western Iberia, 300-600 CE.* Philadelphia: University of Pennsylvania Press.

Fierro, María Isabel. 2007. "The mobile minbar in Cordoba: how the Umayyads of al-Andalus claimed inheritance from the Profet." *Jerusalem Studies in Arabic and Islam* 33: 149-68.

Filipe, Iola, and Carlos Fabião. 2006/2007. "Uma unidade de produção de preparados de peixe de época romana na Casa do Governador da Torre de Bélem (Lisboa): uma primeira apesentação." *Arqueologia & História* 58/59: 103-18.

Fletcher Valls, Domingo. 1958. *Acueductos romanos en el término de Ribarroja del Turia (Valencia).* Valencia: Diputación de Valencia.

Flohr, Miko. 2006. "Organizing the Workshop. Water Management in Roman Fullonicae." In *Cura Aquarum in Ephesus,* edited by Gemma Wiplinger, 193-9. Leuven: Peeters.

Fontes, Luis. 1995. "A igreja sueva de Dume (Braga)." In *IV Reunió d'arqueologia cristiana hispànica,* edited by Josep Maria Gurt Esparraguera and Núria Tena, 415-28. Barcelona: Institut d'Estudis Catalans.

Fontes, Luis, Manuela Martins, Maria do Carmo Ribeiro, and Helena Paula Carvalho. 2010. "A cidade de Braga o seu território nos séculos V-VIII." In *Espacios urbanos en el occidente mediterráneo, ss. VI-VIII,* edited by Alfonso García, 255-62. Toledo: Toletum Visigodo.

Fortes, Mário Luis Soares. 2009. *A xestión da auga na paisaxe romana do occidente peninsular.* PhD thesis, Santiago de Compostela: University of Santiago.

Fuentes Domínguez, Ángel. 2000. "Las termas en la Antigüedad Tardía: reconversión, amortización, desaparición. El caso hispano." In *Termas romanas en el occidente del imperio. II Coloquio internacional de arqueología en Gijón,* edited by Carmen Fernández Ochoa and Virginia García-Entero, 135-46. Gijón: Gijón City Council.

Fuertes Santos, María del Camino, Inmaculada Carrasco Gómez, and Rafael Hidalgo Prieto. 2013. "Una nueva campaña de excavación arqueológica en el palacio de Cercadilla, Córdoba. La secuencia estratigráfica del conjunto termal." *Antiquitas* 25: 137-64.

Gamo Pazos, Emilio, Javier Fernández Ortea, and Jerónimo Sánchez Velasco. 2017. "El abastecimiento de agua a la ciudad iberoromana del cerro de la Virgen de la Muela (Driebes, Guadalajara)." *Veleia* 34: 237-247.

García García, Miguel Ángel. 2007. "Aqua Hispalensis. Primer avance sobre la excavación de la cisterna romana de Plaza de la Pescadería (Sevilla)." *Romula* 6: 125-42.

García Sanjuan, Alejandro. 2016. "La persistencia del discurso nacionalcatólico sobre el Medievo peninsular en la historiografía española actual." *Historiografías: revista de historia y teoría* 12: 132-153.

García Vargas, Enrique. 2012. "La Sevilla tardoantigua. Diez años después (2000-2010)." In *Hispaniae urbes. Investigaciones arqueológicas en ciudades históricas*, edited by José Beltrán Fortes and Oliva Rodríguez Gutiérrez, 881–926. Seville: Universidad de Sevilla.

García-Entero, Virginia. 2005. *Los balnea domésticos –ámbito rural y urbano- en la Hispania romana.* Anejos de Archivo Español de Arqueología 37. Madrid: CSIC.

GEOS. 2010. "El acueducto de Sevilla desde Alcalá de Guadaira." In *Las técnicas y las construcciones en la ingeniería romana*, 205-24. Madrid: Fundación de la Ingeniería Técnica de Obras Públicas.

Giner Iranzo, Daniel. 2007. *Memòria de la intervenció arqueologica preventiva al c/ Coronel Monasterio 6-16 (Barcelona, el Barcelonès). Març-agost de 2004.* Excavation report, Barcelona: MuHBa.

Glick, Thomas. 1970. *Irrigation and society in medieval Valencia.* Cambridge, Mass.: Harvard University Press.

Glick, Thomas, and Helena Kirchner. 2000. "Hydraulic systems and technologies in Islamic Spain: History and archaeology." In *Working with Water in Medieval Europe. Technology and Resource-use*, edited by Paolo Squatriti, 267-330. Leiden: Brill.

Gómez Becerra, Antonio. 1995. "Almuñécar en el tránsito de la Antigüedad a la Edad Media." *Florentia Iliberritana* 6: 175-201.

Gómez de la Torre-Verdejo, Amaya. 2008. "La muralla de Recópolis." In *Recópolis y la ciudad en la época visigoda.* Zona Arqueológica 9, edited by Lauro Olmo Enciso, 77-86. Alcalá de Henares: Museo Arqueológico Regional.

Gómez de Segura, Mercedes, Pedro Dámaso Sánchez Barrero, Nuria Sánchez Capote, and Isaac Sastre de Diego. 2011. "Las conducciones romanas de Mérida. Nuevos datos para su conocimiento." In *Aquam perducendam curavit. Captación, uso y administración del agua en las ciudades de la Bética y el occidente romano*, edited by Lázaro Lagóstena Barrios, José Luis Cañizar Palacios and Lluís Pons Pujol, 129-46. Cádiz: University of Cádiz.

Gómez de Somorrostro, Andrés. 1820. *El acueducto y otras antigüedades de Segovia.* Madrid: Miguel de Burgos.

González Acuña, Daniel. 2011. *Forma Urbis Hispalensis. El urbanismo de la ciudad romana de Hispalis a través de los testimonios arqueológicos.* Seville: University of Seville.

González Fernández, Enrique, and María Covadonga Carreño Gascón. 2007. "Las puertas romanas de la muralla de Lugo." In *Murallas de ciudades romanas en el occidente del imperio. Lucus Augusti como paradigma*, edited by Antonio Rodríguez Colmenero and Isabel Rodá de Llanza, 283-312. Lugo: Diputación Provincial de Lugo.

González Román, Cristobal. 2011. "El agua en las ciudades de la Bética: organización y funciones." In *Aquam perducendam curavit. Captación, uso y administración del agua en las ciudades de la Bética y el occidente romano*, edited by Lázaro Lagóstena Barrios, José Luis Cañizar Palacios and Lluís Lluís Pons Pujol, 41–66. Cádiz: University of Cádiz.

González Tascón, Ignacio, Ana Vázquez de la Cueva, and Antonio Ramírez Sádaba. 1994. *El acueducto romano de Caesaraugusta: según el manuscrito de Juan Antonio Fernández (1752-1814).* Madrid: CEHOPU.

González Tascón, Ignacio, and Isabel Bestué Cardiel. 2006. *Breve guía del patrimonio hidráulico de Andalucía.* Seville: Junta de Andalucía.

Gozalbes Cravioto, Enrique. 2000. "Descubrimientos arqueológicos de Tingi (Tánger) en los siglos X al XVII." *L'Africa Romana* 13.1: 835-52.

Greenhalgh, Michael. 1989. *The Survival of Roman Antiquities in the Middle Ages*. London: Duckworth.

Grewe, Klaus. 1991. "Wasserversorgung und –entsorgung im Mittelalter, ein technikgeschichtlicher Überblick." In *Der Wasserversorgung im Mittelalter*, 11–88. Mainz: Philip von Zabern.

Guidi Sánchez, José Javier. 2016. *De Tàrraco a Terrachona. De la Tarragoba visigoda a la medieval (segles VII-XII): Els condicionants clàssics en la nova arquitectura feudal. Documentació arqueològica i fonts històriques d'un patrimoni monumental.* PhD thesis, Tarragona: Rovira i Virgili University.

Guillén Fernández, Celia. 2003. *Intervención arqueológica en el solar situado en calle Caballeros esquina calle Landerer de Valencia.* Excavation report, Valencia: SIAM.

Gurt Esparraguera, Josep Maria, and Pilar Diarte Blasco. 2012. "La basílica de santa Leocadia y el final de uso del circo romano de Toledo: Una nueva interpretación." *Zephyrus* 69: 149-63.

Gutiérrez Lloret, Sonia. 2012. *El reconocimiento arqueológico de la islamización. Una mirada desde al-Andalus.* Vol. I, in *711. Arqueología e Historia entre dos mundos.* Zona Arqueológica 15, 191–212. Alcalá de Henares: Museo Arqueológico Regional.

Gutiérrez Lloret, Sonia, Lorenzo Abad Casal, and Blanca Gamo Parras. 2005. "Eio, Iyyuh y el Tolmo de Minateda (Hellín, Albacete): de sede episcopal a madîna islámica." In *VI reunió d'arqueologia cristiana hispànica. Les ciutats tardoantigues d'Hispania: Cristianització i topografia*, edited by Josep Maria Gurt Esparraguera and Albert Ribera Lacomba, 345-70. Barcelona: Institut d'Estudis Catalans.

Gutiérrez Lloret, Sonia, and Víctor Cañavate Castejón. 2010. "Casas y cosas: espacios y funcionalidad en las viviendas emirales del Tolmo de Minateda (Hellín, Albacete)." *Cuadernos de Madīnat al-Zahrā* 7: 123-48.

Guyon, Jean. 2006. "La topographie chrétienne des villes de la Gaulle." In *Die Stadt in der Spätantike – Niedergang oder*

Wandel?, edited by Jens-Uwe Krause and Christian Witschel, 106-28. Stuttgart: Steiner.

Haase, Claus-Peter. 1996. "Madinat al-Far: the Regional Late Antique Tradition of an Early Islamic Foundation." In *Continuity and Change in Northern Mesopotamia From the Hellenistic to the Early Islamic Period*, edited by Karin Bartl and Stefan Hauser, 165-72. Berlin: DRV.

Halsall, Guy. 1996. "Towns, societies and ideas: The not-so-strange case of Late Roman and Early Merovingian Metz." In *Towns in Transition: Urban Evolution in Late Antiquity and the Early Middle Ages*, edited by Neil Christie and Simon. Loseby, 235-61. Aldershot: Scholar Press.

Henning, Joachim, Michael McCormick, Lauro Olmo Enciso, Knut Rassman, and Eyub Fikrit. 2019. "Reccopolis revealed: First geomagnetic mapping of the early medieval Visigothic royal town." *Antiquity* 93.369: 735-51.

Hernández Carretero, Ana María. 2003. "Excavación de un tramo de la conducción hidráulica de Cornalvo y nuevas aportaciones al conocimiento de la secuencia ocupacional en la zona de Bodegones." *Mérida, excavaciones arqueológicas* 6: 37-56.

Herreros Hernández, Agustina. 1996. *Intervención arqueológica c/ Quart, 4 y 6.* Excavation report, Valencia: SIAM.

Hidalgo Prieto, Rafael. 1996. "Sobre la interpretación de las termas de Cercadilla (Córdoba)." *Habis* 27: 189-204.

—. 2002. "De edificio imperial a complejo de culto: la ocupación cristiana del palacio de Cercadilla." In *Espacio y usos funerarios en el occidente romano*, edited by Desiderio Vaquerizo Gil, 343-72. Córdoba: University of Córdoba.

—. 2005. "Algunas cuestiones sobre la Corduba de la antigüedad tardía." In *VI Reunió d'arqueologia cristiana hispànica. Les ciutats tardoantigues d'Hispania: Cristianització i topografia*, edited by Josep Maria Gurt Esparraguera and Albert Ribera Lacomba, 401–14. Barcelona: Institut d'Estudis Catalans.

Hiernard, Jean, and José María Álvarez Martínez. 1982. "Aqua Augusta, una inscripción de letras de bronce en Mérida." *Sautola* 3: 221–9.

Hill, Donald. 1993. *Islamic Science and Engineering.* Edinburgh: Edinburgh University Press.

Hodge, Trevor. 1989. "Aqueducts." In *Roman Public Buildings.* Exeter Studies in History 20, edited by Ian Barton, 127-50. Exeter: Exeter University Press.

—. 1992. *Roman Aqueducts and Water Supply.* London: Duckworth.

—. 2000a. "Aqueducts." In *Handbook of Ancient Water Technology*, edited by Örjan Wikander, 39-66. Leiden: Brill.

—. 2002b. "Qanats." In *Handbook of Ancient Water Technology*, edited by Örjan Wikander, 35-8. Leiden: Brill.

Hortelano Uceda, Ignacio. 2008. "La red de acueductos de la Valentia romana. Canales de abastecimiento rural al sur del Turia." *Lucentum* 27: 69-85.

Houten, Pieter. 2018. *Civitates Hispaniae. Urbanisation on the Iberian Peninsula during the High Empire.* PhD thesis, Leiden: University of Leiden.

Ivanišević, Vujadin. 2012. "Akveduct Caričinog Grada – Justinujane Prime." *Saopštenja* 44: 13-31.

—. 2017. "Main patterns of urbanism in Caričin Grad (Justiniana Prima)." In *New Cities in Late Antiquity. Documents and Archaeology*. Bibliothèque de l'Antiquité Tardive 35, edited by Efthymios Rizos, 221–32. Turnhout: Brepols.

Izquierdo Benito, Ricardo. 2009. "¿De complejo palatino visigodo a arrabal islámico?" In *La Vega Baja de Toledo*, edited by Mari Mar Gallego García, Javier García González, Ricardo Izquierdo Benito, Jorge de Juan Ares, Lauro Olmo Enciso, Diego Peris Sánchez and Ramón Villa, 95-113. Toledo: Toletum Visigodo.

Jiménez Sánchez, Juan Antonio, and Jordina Sales Carbonell. 2004. "Termas e iglesias durante la Antigüedad Tardía: ¿reutilización arquitectónica o conflicto religioso? Algunos ejemplos hispanos." In *Sacralidad y arqueología.* Antigüedad y Cristianismo 21, edited by José María Blázquez Martínez and Antonio González Blanco, 185-202. Murcia: University of Murcia.

Jiménez Sancho, Álvaro. 2011. "Itálica. La red de alcantarillado." In *La gestión de los residuos urbanos en Hispania.* Anejos de Archivo Español de Arqueología 60, edited by Josep Anton Remolà Vallverdú and Jesús Acero Pérez, 145-56. Mérida: CSIC.

Jiménez, Alfonso. 1974. "Los acueductos de Bellone Claudia (Bolonia, Cádiz)." *Habis* 5: 273-93.

Jurado Jiménez, Francisco. 2001. "El acueducto romano de Segovia." *Obra pública, ingeniería y territorio* 57.

Keenan-Jones, Duncan, John Hellstrom, and Russell Drysdale. 2008. "Trace element and other analyses of tufa from ancient water systems in Campania and Petra." In *Proceedings of the 13th International Conference on the History of Water Management and Hydraulic Engineering in the Mediterranean Region Petra/Amman 31 March-09 April 2007*, edited by Christoph Ohlig, 329-39. Siegburg: DWhG.

Kelly, Amanda. 2006. "The impact of constructing aqueducts on the settlement patterns of Roman Crete." In *Cura Aquarum in Ephesus*, edited by Gemma Wiplinge, 303-11. Leuven: Peeters.

Kennedy, Hugh. 1998. "From Antiquity to Islam in the cities of al-Andalus and al-Mashriq." In *Genèse de la ville islamique en al-Andalus et au Maghreb occidental*, edited by Patrice Cressier and Mercedes García Arenal, 53-64. Madrid: Casa de Velázquez.

Kulikowski, Michael. 2004. *Late Roman Spain and its Cities.* London: Johns Hopkins University Press.

Lagóstena Barrios, Lázaro. 2001. *La producción de salsas y conservas de pescado en la Hispania Romana (II a.C. - VI d.C.).* Barcelona: University of Barcelona.

Lamprecht, Heinz. 1987. *Opus Caementitium. Bautechnik der Römer.* Düsseldorf: Beton-Verlag.

Lara Medina, Macarena. 2018. "Sobre el abastecimiento, la distribución y la evaciación hídrica de Gades." *Zephyrus* 81: 141–63.

Laurence, Ray, Simon Esmonde Cleary, and Gareth Sears. 2011. *The City in the Roman West c. 250 BC – c. AD 250.* Cambridge: Cambridge University Press.

Lebecq, Stephane. 1996. "Le devenir économique de la cité dans la Gaule des Ve-IXe siècles." In *La fin de la cité antique et le début de la cité médiévale*, edited by Claude Lepelly, 287-310. Bari: Edipuglia.

Lefevbre, Claude. 1997. "L'aqueduc de Gorz à Metz." In *Les aqueducs de la Gaulle romaine et des les regions voisines*. Caesarodunum 31, edited by Robert Bedon, 405-42. Limoges: Presses universitaires de Limoges.

Lemos, Francisco, Manuela Martins, Luis Fontes, Jose Manuel Leite, and Armandinho Cunha. 1998. "A redescoberta da muralha romana e suevica-visigotica de Braga." *Forum* 24: 11–26.

Lenoir, Eliane. 1992. "Enceintes urbaines et thermes de Lixus." In *Lixus. Actes du colloque de Larache (8-11 novembre 1989)*. Collection de l'École Française de Rome 166, 289-98. Rome: École Française de Rome.

—. 2005. "La ville romain de Zilil du Ier siècle av. J.-C. au IVe siècle ap. J.-C." *Pallas* 68: 65-76.

Lévi-Provençal, Évariste. 1957. *España musulmana hasta la caída del califato de Córdoba (711–1031 de J. C.). Instituciones y vida social e intelectual.* Historia de España de Ramón Menéndez Pidal. Vol. V. Madrid: Espasa-Calpe.

Liebeschuetz, Wolfgang. 1992. "The end of the ancient city." In *The City in Late Antiquity*, edited by John Rich, 1–50. London: Routledge.

Linage Conde, Antonio. 1972. "Tras las huellas de Justiniano de Valencia." *Hispania Antiqua* 2: 203-16.

López Quiroga, Jorge. 2004. *El final de la antigüedad tardía en Gallaecia. La transformación de las estructuras de poblamiento entre Miño y Duero (siglos V al X)*. Santiago de Compostela: Fundación Pedro Barrié de la Maza.

López Vilar, Jordi. 2006. *Les basiliques paleocristianes del suburbi occidental de Tàrraco. El temple septentrional i el complex martirial de Sant Fructuós*. Tarragona: ICAC.

López, Adrián, Santiago Domínguez-Solera, and Míchel Muñoz. 2014. "El puente y el acueducto de Valeria." *Ricoti* 19: 81–5.

Loseby, Simon. 2006. "Decline and change in the cities of late antique Gaul." In *Die Stadt in der Spätantike – Niedergang oder Wandel?*, edited by Jens-Uwe Krause and Christian Witschel, 96-104. Stuttgart: Steiner.

Löx, Markus. 2017. "The urban centre of Valencia in late Roman and Visigothic times." *Journal of Late Antiquity* 10.1: 131–65.

Machancoses López, Mirella. 2015. *Topografía Urbana de la Valentia romana altoimperial: Ciudad y Suburbio.* thesis, Valencia: University of Valencia.

Macias Solé, Josep Maria. 2004. *Les termes publiques de l'area portuaria de Tàrraco. Carrer de Sant Miquel de Tarragona.* Tarragona: ICAC.

—. 2008. "Tarracona visigoda ¿una ciudad en declive?" In *Recópolis y la ciudad en la época visigoda.* Zona arqueológica 9, edited by Lauro Olmo Enciso, 293-301. Alcalá de Henares: Museo Arqueológico Regional.

Macias Solé, Josep Maria, and Josep Antón Remolà Vallverdú. 2005. "El port de Tarraco a l'antiguitat tardana." In *VI reunió d'arqueologia cristiana hispànica. Les ciutats tardoantigues d'Hispania: Cristianització i topografia*, edited by Josep Maria Gurt Esparraguera and Albert Ribera Lacomba, 175-87. Barcelona: Institut d'Estudis Catalans.

Macias Solé, Josep Maria, Ignasi Fiz Fernández, Lluís Piñol Masgoret, María Teresa Miró Alaix, and Josep Guitart. 2007. *Planimetria arqueològica de Tàrraco.* Tarragona: ICAC.

Madani, Tariq. 2003 . *L'eau dans le monde musulman médiéval. L'exemple de Fès (Maroc) et de sa région.* PhD thesis, Lyons: Université Lumière Lyon 2.

Magnusson, Roberta, and Paolo Squatriti. 2000. "Technologies of Water Use in Medieval Italy." In *Working With Water in Medieval Europe*, edited by Paolo Squatriti, 217-265. Leiden: Brill.

Manacorda, Danielle. 2001. *Crypta Balbi: Archeologia e storia di un paesaggio urbano.* Rome: Archaeological Superintendency of Rome.

Manzano Moreno, Eduardo. 2006. *Conquistadores, emires y califas. Los Omeyas y la formación de Al-Ándalus.* Barcelona: Crítica.

Marano, Yuri. 2015. "Watered... with the Life-Giving Wave'. Aqueducts and Water Management in Ostrogothic Italy." In *Ownership and Exploitation of Land and Natural Resources in the Roman World*, edited by Paul Erdkamp, Koenraad Verboven and Arjan Zuiderhoek, 150-69. Oxford: Oxford University Press.

Marfil Ruíz, Pedro. 2000a. "La sede episcopal cordobesa en época bizantina: evidencia arqueológica." In *V Reunió d'arqueologia cristiana hispànica*, edited by Josep Maria Gurt Esparraguera and Núria Tena, 157-75. Barcelona: Institut d'Estudis Catalans.

—. 2000b. "Córdoba de Teodosio a Abd al-Rahmán III." In *Visigodos y Omeyas. Un debate entre la Antigüedad tardía y la alta Edad Media (Mérida, abril de 1999).* Anejos de Archivo Español de Arqueología 23, edited by Luis Caballero Zoreda and Pedro Mateos Cruz, 117-42. Madrid: CSIC.

Marín Jordá, Carmen, and Albert Ribera Lacomba. 2000. "Un caso precoz de edificio termal: los baños republicanos de Valentia." In *Termas romanas en el occidente del imperio. II Coloquio internacional de arqueología en Gijón*, edited by Carmen Fernandez Ochoa and Virginia García-Entero, 151–6. Gijón: Gijón City Council.

Márquez Moreno, Carlos, and Ángel Ventura Villanueva. 2005. "Corduba tras las guerras civiles." In *Julio César y Corduba: tiempo y espacio en la campaña de Munda (49-45 a.C.)*, edited by Enrique Melchor Gil, Joaquín Mellado Rodríguez and Juan Francisco Rodríguez-Neila, 429-66. Córdoba: University of Córdoba.

Martí Matías, Miquel. 2001. *Visigodos, hispano-romanos y bizantinos en la zona valenciana en el siglo VI (España).* BAR IS 943. Oxford: Archaeopress.

Martí, Javier, and Josefa Pascual. 2000. "El desarrollo urbano de Madina Balansiya hasta el final del califato." In *Ciudad y territorio en al-Ándalus*, edited by Lorenzo Cara Barrionuevo, 500-38. Granada: Athos-Pregamos.

Martín Viso, Iñaki. 2013. "Prácticas locales de la fiscalidad en el Reino de Toledo." In *Lo que vino de Oriente. Horizontes, praxis y dimensión material de los sistemas de dominación fiscal en Al-Andalus (ss. VII-IX).* BAR IS 2525, edited by Xavier Ballestín Navarro and Ernesto Pastor Díaz, 72-85. Oxford: Archaeopress.

Martínez Jiménez, Javier. 2010. "Uso y suministro de agua a la ciudad de Roma en el periodo Ostrogodo: 476-552 dC." In *Espacios urbanos en el occidente mediterráneo, ss. VI-VIII*, edited by Alfonso García, 267-76. Toledo: Toletum Visigodo.

—. 2011. "The continuity of Roman water supply systems in post-Roman Spain: the case of Valentia, a reliable example?" *Arkeogazte* 1: 125-44.

—. 2012. "Reuse, repair and reconstruction: Functioning aqueducts in post-Roman Spain." In *Make-do and Mend: the Archaeologies of Compromise.* BAR IS 2408, edited by Ben Jervis and Alisn Kyle, 26-44. Oxford: Archaeopress.

—. 2013. "Crisis or crises? The end of Roman towns in Iberia between the late Roman and the early Umayyad periods." In *Tough Times: The Archaeology of Crisis and Recovery. Proceedings of the GAO annual conferences 2010 and 2011*, BAR IS 2478, edited by Elsbeth van der Wilt and Javier Martínez Jiménez, 77-90. Oxford: Archaeopress.

—. 2015. "A preliminary study on the aqueduct of Reccopolis." *Oxford Journal of Archaeology* 34.3: 301–20.

—. 2017. "Water supply in the Visigothic urban foundations of Eio (El Tolmo de Minateda) and Reccopolis." In *New Cities in Late Antiquity. Documents and Archaeology.* Bibliothèque de l'Antiquité Tardive 35, edited by Efthymios Rizos, 233-45. Turnhout: Brepols.

—. 2019. "Tantam pecuniam male perdiderunt: aqueducts and municipal investment in Hispania." In *Signs of Weakness and Crisis in the Western Cities of the Roman Empire (c. II-III AD).* Potsdamer Altertumwissenschaftliche Beiträge 68, edited by Javier Andreu and Aitor Blanco-Pérez, 59-70, Stuttgart: Franz Steiner Verlag.

Martínez Jiménez, Javier, and José María Moreno Narganes. 2015. "Nunc autem a Gothis subversa: The province of Alicante and the Spanish Mediterranean towns between the Byzantine and Visigothic periods." *Early Medieval Europe* 23.3: 263-89.

Martínez Jiménez, Javier, and Patricia González Gutiérrez. 2017. "Knowledge and specialised trades in the late antique West: Medicine vs. Engineering." *Journal for Late Antique Religion and Culture* 11: 38-58.

Martínez Jiménez, Javier, Isaac Sastre de Diego, and Carlos Tejerizo García. 2018. *The Iberian Peninsula 300-850: An Archaeological Perspective.* Late Antique and Early Medieval Iberia 6. Amsterdam: Amsterdam University Press.

Martínez Santa-Olalla, Julio. 1931–1932. "Antigüedades romanas de Poza de la Sal (Burgos)." *Anuario de Prehistoria Madrileña* 2-3: 125-75.

Martins, Manuela. 2005. *As termas romanas do Alto da Cividade. Um exemplo de arquitectura pública em Bracara Augusta.* Bracara Auguta - excavações arqueológicas 1. Braga: University of Braga.

Martins, Manuela, and Luís Fontes. 2010. "Bracara Augusta. Balanço de 30 anos de investigaçâo arqueológica na capital da Galécia Romana." In *Simulacra Romae II: Rome, les capitales de province (capita prouinciarum) et la création d'un espace commun européen. Une approche archéologique*, edited by Ricardo González Villaescusa and Joaquín Ruíz de Arbulo, 111–24. Reims: Société archéologique champenoise.

Martins, Manuela, and Maria do Carmo Franco Ribeiro. 2012. "Gestão e uso da água em Bracara Augusta. Uma abordagem preliminar." In *Caminhos da agua. Paisagens e usos na longa duração*, edited by Manuela Martins, Isabel Vaz de Freitas and María Isabel del Val Valdivieso, 9-52. Braga: Centro de Investigação Transdisciplinar «Cultura, Espaço e Memória».

Martins, Manuela, Fernanda Magalhães, Raquel Peñin, and Jorge Ribeiro. 2016. "The housing evolution of Braga between Late Antiquity and the Early Middle Ages." In *Arqueologia Medieval: Hàbitats Medievals*, edited by Flocel Sabaté and Jesús Brufal, 35-52. Lérida: Pages.

Martins, Manuela, Ricardo Mar, Jorge Ribeiro, and Fernanda Magalhães. 2013. "A construção do teatro romano de Bracara Augusta." In *Historia da construção. Arquitetura e tecnicas construtivas*, edited by Arnaldo Sousa Melo and Maria do Carmo Franco Ribero, 41–74. Braga: Centro de Investigação Transdisciplinar «Cultura, Espaço e Memória».

Mascarenhas, Jose Manuel, Francisco Bilous, and Nuno Sousa Neves. 2012. "O aqueduto romano de Olisipo: viabilidade ou utopia? Ensaio de traçado apoiado em modelação geográfica." *Revista Portuguesa de História* 43: 239-64.

Mateos Cruz, Pedro, Rocío Ayerbe, Tomás Barrientos, and Santiago Feijoo Martínez. 2002. "La gestión del agua en Augusta Emerita." *Empúries* 53: 67-88.

Matthews, John. 1990 [1975]. *Western Aristocracies and Imperial Court, AD 364-425*. Oxford: Oxford University Press.

Mayer Olivé, Marc, and Isabel Rodà de Llanza. 1977. "El abastecimiento de aguas a la Barcelona romana. Reconstrucción de su trazado." In *Segovia. Symposium de arqueología romana*, 265-78. Barcelona: University of Barcelona.

Mayerson, Philip. 1986. "Choricius of Gaza and the water supply system of Caesarea." *Israel Exploration Journal* 36.4: 269-72.

Meiggs, Russell. 1982. *Trees and Timber in the Ancient Mediterranean World*. Oxford: Clarendon.

Meijide Cameselle, Gonzalo, and Francisco Herves Raigoso. 2000. "Un nuevo espacio en las termas de Lugo." In *Termas romanas en el occidente del imperio. II Coloquio internacional de arqueología en Gijón*, edited by Carmen Fernández Ochoa and Virginia García-Entero, 215-20. Gijón: Ayuntamiento de Gijón.

Miller, Maureen. 2000. *The Bishop's Palace. Architecture and Authority in Medieval Italy*. London: Cornell University Press.

Miró Alaix, Carme, and Héctor Orengo. 2010. "El cicle de l'aigua a Barcino. Una reflexió en torn de les noves dades arqueològiques." *Quaderns d'Arqueologia i Historia de la Ciutat de Barcelona* 6: 109-33.

Montalbano, Calogero. 2008. "Dar-al ma: the architecture of water in the Islamic countries." In *The City in the Islamic World*, edited by Salma Jayyusi, 679-730. Leiden: Brill.

Morais, Rui. 2011. "A captação e o uso da água em Bracara Augusta." In *Aquam perducendam curavit. Captación, uso y administración del agua en las ciudades de la Bética y el occidente romano*, edited by Lázaro Lagóstena Barrios, José Luis Cañizar Palacios and Lluís Pons Pujol, 225-43. Cádiz: University of Cádiz.

Moreno Martín, Francisco. 2018. "Circulación de modelos y circularidad de argumentación para el conocimiento de la topografía cristiana entre la Antigüedad Tardía y la Alta Edad Media." In *Entre civitas y madina. El mundo de las ciudades en la Península Ibérica y en el norte de África (siglos IV-IX)*, Collection de la Casa de Velázquez 167, edited by Sabine Panzram and Laurent Callegarin, 153-72. Madrid: Casa de Velázquez.

Moreno Rosa, Antonio, and Guadalupe Pizarro Berengena. 2011. "La continuidad de los sistemas hidráulicos. Nuevos testimonios en Córdoba." In *Aquam perducendam curavit. Captación, uso y administración del agua en las ciudades de la Bética y el occidente romano*, edited by Lázaro Lagóstena Barrios, José Luis Cañizar Palacios and Lluís Pons Pujo, 165-82. Cádiz: University of Cádiz.

Mueden, Rachid. 2010. *Las colonias y municipios de la Mauretania Tingitana.* PhD thesis, Granada: University of Granada.

Murillo Redondo, Juan Francisco. 2009. "La almunia de al-Rusafa en Córdoba." *Madrider Mitteliungen* 50.2: 449-82.

Murillo Redondo, Juan Francisco, José Ramón Carrillo, Maudilio Moreno, Dolores Ruíz, and Sonia Vargas. 2002. "Los monumentos funerarios de Puerta de Gallegos- Colonia Patricia Corduba." In *Espacio y usos funerarios en el occidente romano*, edited by Desiderio Vaquerizo Gil. Córdoba: University of Córdoba.

Navarro Palazón, Julio, and Pedro Jiménez Castillo. 1995. "El agua en la vivienda andalusí. Abastecimiento, almacenamiento y evacuación." *Verdelay* 7: 401–12.

Nielsen, Inge. 1990. *Thermae et Balnea. The Architecture and Cultural History of Roman Public Baths.* Aarhus: Aarhus University Press.

Ohlig, Christoph. 2004. "'Ein castellum aquae ist wie jedes andere, und alle sind gebaut nach den Prinzipien des Vitruv' – Zwei offensichtlich nur schwer auszurottende Irrtümer." In *Wasserbauten im Köningreich Urartu und weitere Beiträge zur Hydrotechnik in der Antike*, edited by Christoph Ohlig, 133-81. Siegburg: DWhG.

Oller Guzmán, Joan. 2014. "La civitas sine urbe y su funciónalidad de vertebración en el territorio provincial hispano: los casos de Egara y de Caldes de Montbui." *Pyrenae* 45.1: 89-110.

Olmo Enciso, Lauro. 2006. *Recópolis: un paseo por la ciudad visigoda.* Alcalá de Henares: Museo Arqueológico Regional.

—. 2008. "Recópolis: una ciudad en época de transformaciones." In *Recópolis y la ciudad en la época visigoda.* Zona Arqueológica 9, edited by Lauro Olmo Enciso, 40-63. Alcalá de Henares: Museo Arqueológico Regional.

—. 2009. "La Vega Baja en época visigoda: una investigación arqueológica en construcción." In *La Vega Baja de Toledo*, edited by Mari Mar Gallego García, Javier García González, Ricardo Izquierdo Benito, Jorge de Juan Ares, Lauro Olmo Enciso, Diego Peris Sánchez and Ramón Villa, 69-94. Toledo: Toletum Visigodo.

Olmo Enciso, Lauro, Manuel Castro Priego, Amaya Gómez de la Torre-Verdejo, and Álvaro Sanz Paratcha. 2008. "Recópolis y su justificación científica: la secuencia estratigráfica." In *Recópolis y la ciudad en la época visigoda.* Zona Arqueológica 9, edited by Lauro Olmo Enciso, 64-75. Alcalá de Henares: Museo Arqueológico Regional.

Orengo, Héctor, and Carme Miró Alaix. 2011. "Following Roman waterways from a computer screen. GIS-based approaches to the analysis of Barcino's aqueducts." In *Go Your Own Least Cost Path. Proceedings of the GIS session at EAA 2009. Riva del Garda.* BAR IS 2284, edited by Axel Posluschny, Alžběta Danielisová Philip Verhagen, 47-53. Oxford: Archaeopress.

Osland, Daniel. 2011. *Urban Change in Late Antique Hispania: The Case of Augusta Emerita.* PhD Thesis, Cincinnati: University of Cincinnati.

Pallarés Salvador, Francisco. 1969. "Las excavaciones en la plaza de San Miguel y la topografía romana de Barcino." *Cuadernos de Arqueología e Historia* 13: 5-42.

Palol Solelles, Pedro. 1967. *Arqueología cristiana de la España romana.* Madrid: CSIC.

Panzram, Sabine. 2002. *Stadtbild und Elite: Tarraco, Corduba und Augusta Emerita zwischen Republik und spätantike.* Stuttgart: Franz Steiner Verlag.

Passchier, Cees, Gür Sürmelihindi, Christoph Spötl, Regina Mertz-Kraus, and Denis Scholz. 2016. "Carbonate deposits from the ancient aqueduct of Béziers, France — A high-resolution palaeoenvironmental archive for the Roman Empire." *Palaeogeography, Palaeoclimatology, Palaeoecology* 461: 328-40.

Passini, Jean. 2010. "El baño de Bab al-Mardun (o de la Cruz): localización e identificación." *Al-Qantara* 31.1: 211–23.

Pavón Maldonado, Basilio. 1988. "Entre la historia y la arqueología. El enigma de la Córdoba califal desaparecida (II)." *Al-Qantara* 9.2: 403-25.

—. 1990. *Tratado de arquitectura hispano-musulmana I. Agua.* Madrid: CSIC.

Peral Bejarano, Carmen. 1995. "La infraestructura de aguas urbanas en la Málaga andalusí." In *Agricultura y regadío en al-Andalus. II Coloquio de Historia y Medio Físico*, edited by Antonio Cara Barrionuevo and Antonio Malpica Cuello, 117-32. Almería: Instituto de Estudios Almerienses.

Pérex Agorreta, María Jesús, and Carme Miró Alaix. 2018. *Ubi aquae ibi salus. Aguas mineromedicinales, termas curativas y culto a las aguas en la Península Ibérica (desde la Protohistoria a la Tardoantigüedad).* Madrid: UNED.

Pérez Maestro, Carmen. 2005. "Nuevos datos acerca de la conducción hidráulica de Cornalvo." *Mérida, excavaciones arqueológicas* 8: 269-79.

Pérez Marrero, Jenny. 2011. "Nuevas aportaciones al estudio hidráulico del acueducto romano de Tempul." In *Aquam perducendam curavit. Captación, uso y administración del agua en las ciudades de la Bética y el occidente romano*, edited by Lázaro Lagóstena Barrios, José Luis Cañizar Palacios and Lluís Pons Pujol, 183-95. Cádiz: University of Cádiz.

Pérez Martínez, Meritxell. 2012. *Tarraco en la Antigüedad Tardía. Cristianización y organización eclesiástica (siglos III a VIII).* Tarragona: Arola Editors.

—. 2014. "El final del Imperio romano de Occidente en Tarraco. La inscripción de los emperadores León I y Anthemio (467-472 d.C.)." *Pyrenae* 45.2: 117-38.

Pérez Mínguez, Rafael. 2006. *Aspectos del mundo rural romano en el territorio comprendido entre los ríos Turia y Palancia.* Valencia: Diputación de Valencia.

Perich Roca, Arnau. 2014. "Barcino entre los siglos IV y VI dC: transformaciones y ascenso de una ciudad mediterránea durante la antigüedad tardía." *Espacio, Tiempo y Forma, Serie I* 7: 61–96.

Perich Roca, Arnau, and Ferran Gris Jeremías. 2015. "Las fases tardorromana y visigoda de al "Casa de los Mármoles" (Mérida, España). Análisis arquitectónico y nuevas propuestas de restitución." *Oppidum. Cuadernos de investigación* 11: 171–98.

Pickett, Jordan. 2017. "Water and Empire in the de Aedificiis of Procopius." *Dumbarton Oaks Papers* 71: 95-125.

Pizarro Berengena, Guadalupe. 2012. *El abastecimiento de agua a Córdoba. Arqueología e historia.* PhD thesis, University of Córdoba.

Pons Pujol, Lluis, and Lázaro Lagóstena Barrios. 2011. "Los acueductos de Mauretania Tingitana. Estado de la cuestión." In *Aqvam perdvcendvm cvravit. Captación, uso y administración del agua en las ciudades de la Bética y el occidente romano*, edited by Lázaro Lagóstena Barrios, Cañizar Palacios, José Luis and Lluis Pons Pujol, 533-42. Cádiz: University of Cádiz.

Porath, Yosi. 2002. "Hydraulic plaster in aqueduts as a chronological indicator." In *The Aqueducts of Israel.* JRA Supplement Series

46, edited by Davis Amit, Patrich, Joseph and Yizhar Hirschfeld, 25-36. Portsmouth: JRA.

Porres Martín-Cleto, Julio. 1970. "El abastecimiento romano de aguas a Toledo." In *IV asamblea de instituciones de cultura de las diputaciones*. Toledo: Diputación de Toledo.

Posac Mon, Carlos. 1977. "El acueducto de Los Arcos Quebrados (Ceuta)." In *Segovia. Symposium de arqueología romana*, 325-8. Barcelona: University of Barcelona.

Puig Verdaguer, Ferrán, and Isabel Rodá de Llanza. 2007. "Las murallas de Barcino. Nuevas aportaciones al conocimiento de la evolución de sus sistemas de fortificación." In *Murallas de ciudades romanas en el occidente del imperio. Lucus Augusti como paradigma*, edited by Antonio Rodrígez Colmenero and Isabel Rodá de Llanza, 595-632. Lugo: Diputación de Lugo.

Raffard, Delphine, Marc Vinches, Jean-Pierre Henry, Philippe Leveau, Mireille Goutouch, and Robert Thernot. 2000. "The building of the Roman aqueducts: financial and technological problems. The case of the Arles aqueduct." In *Cura Aquarum in Sicilia*, edited by Gemma Jansen, 125-34. Leuven: Peeters.

Ramallo Asensio, Sebastián, Antonio Murcia Muñoz, and Jaime Vizcaíno Sánchez. 2010. "Carthago Nova y su espacio suburbano. Dinámicas de ocupación en la periferia de la urbs." In *Las áreas suburbanas en la ciudad histórica. Topografía, usos y función*, edited by Desiderio Vaquerizo Gil, 211-54. Córdoba: University of Córdoba.

Ramírez Gallardo, Aurelio. 1975. *Supervivencia de una obra hidráulica. El acueducto de Segovia.* Segovia: Colegio de Ingenieros de Caminos Canales y Puertos.

Rėklaitytė, Ieva. 2012. *Vivir en una ciudad de al-Ándalus. Hidráulica, saneamiento y condiciones de vida.* Monografías arqueológicas 47. Zaragoza: University of Zaragoza.

Remolà Vallverdú, Josep Anton, and Jesús Acero Pérez. 2011. *La gestión de los residuos urbanos en Hispania.* Anejos de Archivo Español de Arqueología 60. Mérida: CSIC.

Remolà Vallverdú, Josep Anton, and Joaquín Ruiz de Arbulo. 2002. "L'aigua a la colònia Tarraco." *Empúries* 53: 29-65.

Retuerce Velasco, Manuel. 2004. "Testimonios materiales del Madrid andalusí." In *Testimonios del Madrid medieval. El Madrid musulmán*, edited by Araceli Turina Gómez, Amalia Pérez Navarro and Salvador Quero Castro, 81–116. Madrid: Museo de San Isidro.

Ribeiro, Maria do Carmo Franco. 2008. *Braga entre a época romana e a Idade Moderna. Uma metodologia de análise para a leitura da evolução da paisagem urbana.* PhD thesis, Braga: University of Minho.

Ribeiro, Ricardo, Nuno Neto, Paulo Rebelo, and Rocha Miguel. 2017. "Dados preliminares de uma intervenção arqueológica nos antigos armazéns Sommer (2014-2015). Três mil anos de história da cidade de Lisboa." In *I encontro de arqueologia de Lisboa: Uma cidade em escavação*, edited by Ana Caessa, Cristina Nozes, Isabel Cameira and Rodrigo Banha da Silva, 223-45. Lisbon: Câmara Municipal de Lisboa.

Ribera Lacomba, Albert. 2005. "Origen i desevolupament del nucli episcopal de València." In *VI Reunió d'arqueologia cristiana hispànica. Les ciutats tardoantigues d'Hispania: Cristianització i topografia*, edited by Josep Maria Gurt Esparraguera and Albert Ribera Lacomba, 207-43. Barcelona: Institut d'Estudis Catalans.

—. 2008. "La ciudad de Valencia durante el periodo visigodo." In *Recópolis y la ciudad en la época visigoda. Zona Arqueológica 9*, edited by Lauro Olmo Enciso, 303-20. Alcalá de Henares: Museo Arqueológico Regional.

Ribera Lacomba, Albert, and Miquel Rosselló Mesquida. 200. "El primer grupo Episcopal de Valencia." In *Los orígenes del cristianismo en Valencia y su entorno*, edited by Albert Ribera Lacomba and Lorenzo Abad Casal, 165-85. Valencia: Ayuntamiento de Valencia.

Riera Frau, M. Magdalena. 1993. *Evolució urbana i topografia de Madîna Mayûrqa.* Palma de Mallorca: Ajuntament de Palma.

Rigal, Didier. 1997. "L'aqueduc de Cahors." In *Les aqueducs de la Gaulle romaine et des les regions voisine*, edited by Robert Bedon, 311–32. Limoges: Presses Universitaires de Limoges.

Ríos González, Sergio. 1997. "Arquitectura del agua en la Alta Edad Media. El ejemplo de Foncalada (Oviedo)." In *Termalismo Antiguo*, edited by María Jesús Peréx Agorreta, 493-501. Madrid: UNED.

Ripoll López, Gisela. 2000. "Sedes regiae en la Hispania de la Antigüedad Tardía." In *Sedes Regiae (ann. 400-800)*, edited by Gisela Ripoll López and Josep Maria Gurt Esparraguera, 371–404. Barcelona: Real Academia de Buenas Letras de Barcelona.

—. 2002. "The transformation of the city of Barcino during Late Antiquity." In *From Barcino to Barchinona (1st to 7th centuries). The Archaeological Remains of the Plaça del Rei in Barcelona*, edited by Julia Beltrán de Heredia, 34-47. Barcelona: Museo de Historia de la Ciudad.

Ripoll López, Gisela, and Isabel Velázquez Soriano. 2008. "Toletum vs. Recópolis. ¿Dos sedes para dos reyes?" In *Recópolis y la ciudad en la época visigoda*. Zona Arqueológica 9, edited by Lauro Olmo Enciso, 204-219. Alcalá de Henares: Museo Arqueológico Regional.

Rodríguez Colmenero, Antonio. 2011. *Lucus Augusti. La ciudad romano-germánica del Finisterre ibérico. Génesis y evolución histórica (14 a.C. – 711 d.C.)*. Lugo: Concello de Lugo.

Rodríguez Colmenero, Antonio, and Isabel Rodà de Llanza. 2007. *Murallas de ciudades romanas en el occidente del Imperio*. Lugo: Museo Provincial de Lugo.

Rodríguez Hidalgo, José Manuel. 1997. "La nueva imagen de la Itálica de Adriano." In *Itálica MMCC*, edited by María Pilar León Alonso and Antonio Caballos Rufino, 87-114. Seville: Junta de Andalucía.

Rodríguez Pascua, M. Ángel, Lorenzo Abad Casal, Raúl Pérez López, Blanca Gamo Parras, Pablo Gabriel Silva Barroso, Víctor Hugo Garduño Monroy, Jorge Giner Robles, et al. 2013. "Roman, Visigothic and Islamic evidence of earthquakes recorded in the archaeological site of "El Tolmo de Minateda" (Prebetic Zone, southeast of Spain)." *Cuaternario y Geomorfología* 27.4: 83-90.

Rosselló Mesquida, Miquel. 1996. "El yacimiento de València la Vella (Riba-roja de Túria, Valencia). Algunas consideraciones para su atribución cronológica y cultural." *Quaderns de Prehistória i Arqueologia de Castelló* 17: 435-55.

Ruiz Bueno, Manuel. 2015. "El kardo maximus de Córdoba en la antigüedad tardía." *Anales de Arqueología Cordobesa* 25-26: 83-114.

—. 2016. *Topografía, imagen y evolución urbanística de la Córdoba clásica a la tardoantigua (ss. II-VII d.C.).* PhD Thesis, Córdoba: University of Córdoba.

—. 2017. "Captación y almacenamiento de agua en el espacio intramuros de Córdoba (siglos III-VII): nuevas reflexiones." *Antesteria* 9: 149-63.

—. 2018a. "La desarticulación del callejero hispanorromano: cambios en la infraestructura viaria y de saneamiento entre los siglos II y VII d. C." *Archivo Español de Arqueología* 91: 143-162.

—. 2018b. *Dinámicas topográficas urbanas en Hispania: el espacio intramuros entre los siglos II y VII dC.* Bari: Edipuglia.

Ruiz Prieto, Enrique. 2013. "Itálica tardoantigua: reflexiones y asignaturas pendientes." *Ligustinus* 1: 81–117.

Sáenz Ridruejo, Fernando. 1977. "Observaciones técnicas sobre el abastecimiento romano de aguas a Tarragona." In *Segovia. Symposium de arqueología romana*, 351–8. Barcelona: University of Barcelona.

Salcedo Hierro, Miguel. 2000. *La mezquita, catedral de Córdoba.* Córdoba: Ayuntamiento de Córdoba.

Salvatierra Cuenca, Vicente, José Luis Serrano Peña, and Carmen Pérez Martínez. 1998. "La formación de la ciuclad en al-Andalus. Elementos para una nueva propuesta." In *Genèse de la ville islamique en al-Andalus et au Maghreb occidental*, edited by Patrice Cressier and Mercedes García Arenal, 185-206. Madrid: Casa de Velázquez.

Sánchez López, Elena. 2018. "The role of fresh water in fish processing in Antiquity." *Journal of Maritime Archaeology* 13.1: 83-96.

Sánchez López, Elena, and Enrique Gozalbes Cravioto. 2012. "Los usos del agua en la Hispania romana." *Vínculos de Historia* 1: 11–29.

Sánchez López, Elena, and Javier Martínez Jiménez. 2016. *Los acueductos de Hispania. Construcción y abandono.* Madrid: Fundación Juanelo Turriano.

Sánchez López, Elena, and Santiago Moreno Pérez. 2012. "El acueducto de Almuñécar. Revisión de su trazado y contextualización." *Pyrenae* 43.1: 83-106.

Sánchez López, Elena, Jeny Pérez Marrero, Margarita Orfila Pons, and Ignacio Bestué Cardiel. 2011. "El municipium Sexi Firmum Iulium y el agua. El acueducto y la producción de salazones." In *Aquam perducendam curavit. Captación, uso y administración del agua en las ciudades de la Betica,* edited by Lázaro Lagóstena Barrios, José Luis Cañizar Palacios and Lluis Pons Pujol, 197-216. Cádiz: University of Cádiz.

Sánchez Ramos, Isabel. 2010. *Corduba durante la Antigüedad Tardía. Las Necrópolis urbanas. BAR IS 2126.* Oxford: Archaeopress.

Sánchez Real, José, Pau Pujante Izquierdo, and Ramón Pau Bages. 1994. "Datos para el estudio de los acueductos romanos de Tarragona." *Quaderns d'Historia Tarraconense* 13: 107-15.

Sánchez Velasco, Jerónimo. 2011. "Corduba." In *La gestión de los residuos urbanos en Hispania.* Anejos de Archivo Español de Arqueología 60, edited by Josep Antón Remolà Vallverdú and Jesús Acero Pérez, 123-44. Mérida: CSIC.

—. 2018. *The Christianization of Western Baetica. Architecture, Power and Religion in a Late Antique Landscape.* Late Antique and Early Medieval Iberia 5. Amsterdam: Amsterdam University Press.

Santillana, Davide. 1926. *Istituzioni di diritto musulmano malichita con riguardo anche al sistema sciafiita.* Rome: ARE.

Sanz Paratcha, Álvaro. 2008. "Vida después de la muerte: los contextos cerámicos de Recópolis en época emiral." In *Recópolis y la ciudad en la época visigoda.* Zona Arqueológica 9, edited by

Lauro Olmo Enciso, 165-79. Alcalá de Henares: Museo Arqueológico Regional.

Saradi, Hélène. 2006. *The Byzantine City in the Sixth Century. Literary Images and Historical Reality.* Athens: Society of Messenian Archaeological Studies.

Sastre de Diego, Isaac. 2013. *Los altares de las iglesias hispanas tardoantiguas y altomedievales.* BAR IS 2503. Oxford: Archaeopress.

—. 2015. *Mérida capital cristiana. De Roma a al-Andalus.* Cuadernos Emeritenses 41. Mérida: MNAR.

Sastre de Diego, Isaac, and Javier Martínez Jiménez. 2012. *Informe de las excavaciones en Casa Herrera. Campaña del XVI Curso Internacional de Arqueología de Mérida (2012).* Excavation report, Mérida: Consorcio de Mérida.

—. 2013. *Informe de las excavaciones en Casa Herrera. Campaña del XVII Curso Internacional de Arqueología de Mérida (2013).* Excavation report, Mérida: Consorcio de Mérida.

Silva Cordero, Andrés. 2003. "Un nuevo tramo de la conducción hidráulica Proserpina-Los Milagros." *Mérida, excavaciones arqueológicas* 6: 289-95.

Silva, Rodrigo Banha da, and Adriaan De Man. 2013. "Palácio dos Condes de Penafiel: A significant late antique context from Lisbon." In *Actas do X Congresso Internacional A Cerâmica Medieval no Mediterrâneo*, 397-402. Mértola: Campo Arqueológico de Mértola.

Siraj, Ahmed. 1994a. "De Tingi à Tandja: le mystère d'une capitale déchue." *Antiquités africaines* 30: 281-302.

—. 1994b. "Tanger à l'époque romaine et arabe." *L'Africa Romana* 10.1: 221–9.

Sol Vallés, Joan. 1977. "Notas sobre unas catas efectuadas en el solar de la casa nº3 de la calle Arcs de Barcelona, intentando situar el acueducto romano." In *Segovia. Symposium de arqueología romana*, 377-82. Barcelona: University of Barcelona.

Solier, Yves. 1991. *La basilique paléochretienne du Clos de la Lombarde a Narbonne. Cadre archéologique, vestiges et mobilier*. Revue Archéologique du Narbonnaise 23. Narbonne: CNRS.

Soriano Sánchez, Rafaela. 2000. "El monumento funerario de la cárcel de San Vicente y las tumbas privilegiadas." In *Los orígenes del cristianismo en Valencia y su entorno*, edited by Albert Ribera Lacomba and Lorenzo Abad Casal, 187-92. Valencia: Ayuntamiento de Valencia.

Souto Lasala, Juan Antonio. 2001. "Arquitectura y estado omeya." In *El esplendor de los Omeyas cordobeses*, edited by Antonio Rodríguez Jiménez, 29-31. Córdoba.

Squatriti, Paolo. 1998. *Water and Society in Early Medieval Italy, AD 400-1000*. Cambridge: Cambridge University Press.

Stambaugh, John. 1988. *The Ancient Roman City*. London: John Hopkins University Press.

Sürmelihindi, Gür, Cees Passchier, Christoph Spötl, Paul Kessener, Michel Bestmann, Borrit Jacob, and Orhan Baykan. 2013. "Laminated carbonate deposits in Roman aqueducts: Origin, Processes and implications." *Sedimentology* 60: 961–82.

Tarradellas Corominas, Cristina. 2000. "Topografía urbana de Sevilla durante la Antigüedad Tardía." In *V Reunió d'arqueologia cristiana hispànica*, edited by Josep Maria Gurt and Nuria Tena, 279-90. Barcelona: Institut d'Estudis Catalans.

Teja Casuso, Ramón, and Silvia Acerbi. 2010. "El palacio visigodo y el circo de Toledo: hipótesis de localización." *Reti Medievali* 11: 81–6.

Tejerizo García, Carlos. 2017. *Arqueología de las sociedades campesinas en la cuenca del Duero durante la Primera Alta Edad Media*. Bilbao: University of the Basque Country.

Tsuki, Tsvika, and Yehuda Peleg. 2017. "The development of water supply systems until the Middle Ages." In *Wasserversorgung in Toledo und Wissenvermittlung von der Antike ins Mittelalter*. Iberia Archaeologica 19, edited by Thomas Schattner and Fernando Valdés, 59-82. Berlin: DAI.

Unzu Urmeneta, Mercedes, Paz Prieto, María Jesús Pérex Agorreta, and Amparo and Hernando Grande. 2006. "Roman Baths in

Pamplona (Navarra, Spain)". In *Cura aquarum in Ephesus*, edited by Gilbert Wiplinger, 431–36. Leuven: Peeters.

Utrero Agudo, María Ángeles. 2006. *Iglesias tardoantiguas y altomedievales en la Península Ibérica. Análisis arqueológico y sistemas de abovedamiento.* Anejos de Archivo Español de Arqueología 40. Madrid: CSIC.

Valdés Fernández, Fernando. 1992. "El Madrid islámico: notas para una división arqueológica." In *Mayrit. Estudios de arqueología medieval madrileña*, edited by Fernando Valdés Fernández, 141–80. Madrid: Polifemo.

Vallejo Triano, Antonio. 2010. *La ciudad califal de Medinat al-Zahra. Arqueología de su arquitectura.* Córdoba: Almuzara.

Valverde Castro, María Rosario. 2000. *Ideología, simbolismo y ejercicio del poder real en la monarquía visigoda: un proceso de cambio.* Salamanca: Ediciones Universidad de Salamanca.

Vaquerizo Gil, Desiderio, and Juan Francisco Murillo Redondo. 2016. "The suburbs of Córdoba." *Estoa* 5.9: 37-60.

Ventura Villanueva, Ángel. 1993. *El abastecimiento de agua a la Córdoba romana I: el acueducto de Valdepuentes.* Córdoba: University of Córdoba.

—. 1996. *El abastecimiento de agua a la Córdoba romana II: acueductos, ciclos de distribución y urbanismo.* Córdoba: University of Córdoba.

—. 2002. "Los acueductos romanos de Córdoba y su rehabilitación omeya." *Empúries* 53: 113-30.

Ventura Villanueva, Ángel, and Guadalupe Pizarro Berengena. 2010. "El Aqua Augusta (acueducto de Valdepuentes) y el abastecimiento de agua a la Colonia Patricia Corduba: investigaciones recientes (2000-2010)." In *Las técnicas y las construcciones en la ingeniería romana*, 175-204. Madrid: Fundación de la Ingeniería Técnica de Obras Públicas.

Vidal Castro, Francisco. 2000. "Agua y urbanismo: evacuación de aguas en fatwà-s de al-Andalus y el Norte de África." In *L'urbanisme dans l'occident musulmana u Moyen Âge. Aspects juridiques*, edited by Patrice Cressier, María Isabel Fierro and Jan-Peeter van Staëuel, 101–23. Madrid: CSIC.

Viñes Pérez, Asunción. 1997. *Informe arqueológico de la intervención realizada en la calle Caballeros nº 26 de Valencia.* Excavation report, Valencia: SIAM.

Vizcaíno Sánchez, Jaime. 2009. *La presencia bizantina en Hispania (siglos VII-VIII). La documentación arqueológica.* Antigüedad y Cristianismo 24. 2nd. Murcia: University of Murcia.

Ward-Perkins, Bryan. 1984. *From Classical Antiquity to the Middle Ages. Urban Public Building in Northern and Central Italy AD 300-850.* Oxford: Oxford University Press.

—. 1997. "Continuists, catastrophists and the towns of post-Roman northern Italy." *Papers of the British School at Rome* 65: 137-76.

—. 2005. *The Fall of Rome and the End of Civilization.* Oxford: Oxford University Press.

Ward-Perkins, John Bryan. 1974. *Cities in Ancient Greece and Italy: Planning in Classical Antiquity.* New York: George Bazillier.

Werther, Lukas, Peter Dietrich, Stefanie Berg-Hobohm, and Peter Ettel. 2017. "Der Kanal Karls des Großen Ein frühmittelalterliches Projekt europäischer Dimension." *Geographische Rundschau* 9: 12-18.

Whittow, Mark. 1990. "Ruling the late Roman and early Byzantine city: a continuous history." *Past and Present* 129: 3-29.

Wickham, Chris. 2005. *Framing the Early Middle Ages. Europe and the Mediterranean 400-800.* Oxford: Oxford University Press.

Wilson, Andrew. 2000. "The water-mills on the Janiculum." *Memoires of the American Academy in Rome* 45: 219-46.

—. 1997. *Water Management and Usage in Roman North Africa: a Social and Technological Study*, DPhil thesis. Oxford: University of Oxford.

—. 2012. "Water, power and culture in the Roman and Byzantine worlds: an introduction." *Water History* 4.1: 1–10.

Wood, Jamie, and Javier Martínez Jiménez. 2016. "New directions in the study of Visigothic Spain." *History Compass* 14.1: 29-38.

Wyss, Michel. 1996. *Atlas historique de Saint-Denis. Des origines au XVIIIe siècle.* Paris: Editions de la maison des sciences de l'homme.

Yegül, Fikret. 1991. *Baths and Bathing in Classical Antiquity.* Cambridge, Mass.: MIT Press.

Zamora Canellada, Alonso. 2007. "Notas sobre el estado actual de las investigaciones arqueológicas del acueducto de Segovia." In *El agua y las ciudades romanas,* edited by Julio Mangas and Santiago Martínez, 133-46. Madrid: Ediciones 2007.

Zozaya, Juan. 2001. "Fortificaciones tempranas en al-Andalus, ss. VIII-X." In *Mil Anos de Fortificações na Península Ibérica e no Magreb (500-1500): Actas do Simpósio Internacional sobre Castelos,* 45-58. Lisbon: Colibri.

General Index

Aachen, 55
Abd al-Rahman I, emir, 79, 114, 119, 120, 137, 163, 173, 239
Abd al-Rahman III, caliph, 70, 117, 163, 166, 174, 234
Aïn-el-Hammam, 254
al-Bakri, 16, 258, 259, 260
al-Idrisi, 16, 83, 139, 140, 201-2
aljibe, see cisterns
Almuñécar, 40, 137-41, **138**
Almuñécar - aqueduct, 17, 39, 46, 87, 129, 138-40, **139, 140**
Ampurias, 8, 85
Andelos, 38
aqueducts, *passim*
aqueducts - damage, 27, 38, 47, 48-50, 68, 132, 141, 202-3, 215, 219
aqueducts - definition, 9-10
aqueducts - historiography, 18
aqueducts - maintenance, 29, 30, 45-8, 72, 98-101, 104, 107, 110-1, 132, 165, 243
aqueducts - new constructions, 39, 51, 104, 107-8, 112, 115, 127-8, 132, 208
aqueducts - quantification, 35-6, 39, 254
aqueducts - symbolism, 2, 11, 33, 93, 120-1, 127, 131
arcuationes and arches, 10, 18, 43, 49, 51, 120, 138-40, 145-6, 151, 159, 163, 164, 165, 167, 171, 183, 188, 189, 190, 192, 201, 208, 211, 214-5, 216-7, 225, 240, 258-9
Arles, 12, 20, 32, 50
Arucci, 38

Baelo Claudia, 18, 26, 38, 49, 50
Banasa, 254, 255, 261, 265
baptisteries, 25, 62, 110, 142, 150, 159, 200, 204, 210, 219, 229, 233, 237, 242, 245
Barbarians (5th c.), see Germanic peoples, Sueves, and Visigoths
Barcelona, 40-1, 50, 75, 79, 101, 141-53, **146**, 222
Barcelona - aqueduct, 16, 39, 46, 88, 144-8, **148**, 150-3
Barcelona - domestic baths, 73, 97, 144, 200-1
Barcelona - episcopal baths, 25, 111, 131, 143, 150-1, **152**
Barcelona - episcopal complex, 46, 75, 79, 90, 110-1, 142-3, **143**, 148, 150, 219, 235
Barcelona - sewers, 28, 149
Basra, 264
Baths and bathing - culture, 12, 24, 92, 97, 99, 101, 112
Baths and bathing - design, 25
Baths and bathing - functioning, 24, 58-9

Baths and bathing - hammams, 126-7, 239, 248-9, 264
Baths and bathing - private, 12, 73, 97, 144, 216, 233, 235, 252
Baths and bathing - public, 12, 99, 177, 233, 236
Baths and bathing - reuse, 85
Begastri, 38
Bilbilis, 8
bishops and the Church, 13, 24, 31, 33, 78, 89, 95, 101, 105, 109-13, 131, 142, 152, 161, 179, 187, 210, 235-7, 249, 256
Braga, 41, 50, 58, 78, 79, 153-7, **154**
Braga - aqueduct, 39, 154-5
Braga - baths, 85, 154-6
Braga - Carvalheiras excavation, 78, 154
burials, cemeteries, tombs, 76, 80, 87-8, 90, 155, 156, 158, 163, 172, 178, 198, 203, 210, 212, 233, 239, 257
Byzantines and the Province of *Spania*, 6, 39, 40, 49, 53, 54, 57, 73, 90, 137, 141, 162, 177, 236, 254, 257, 260, 261, 265

Cádiz, 37
Caparra, 37
caput aquae, 10, 42, 49, 56, 147, 154, 159, 164, 166, 167, 169, 170, 171, 172, 178, 179, 183, 188, 190, 191, 197, 207, 210, 211, 225, 227, 240, 251, 258
Caraca, 38
Cartagena, 11, 27-8, 39, 46, 49, 56-7, **57**, 59, 85, 86
Cartagena - theatre, 19-20, 73-4
Carthage, 12, 254, 263
Casa Herrera, 78, 88, 111, 199-200, **199**
Cassiodorus, 1, 15, 54, 103-4
castellum aquae/divisorium, 10, 20, 28, 29, 38, 39, 43, 79, 81, 85, 87, 96, 140, 149, 155, 159, 169, 170, 172, 183, 190, 195, 217, 228, 232, 235, 242, 245, 251
Castulo, 37
Cercadilla complex, 41, 76, 161-2, 171-2
cesspits, 74, 125
Ceuta, 254, 256-60, **258, 259**, 263, 264
Chaves, 38
Christianisation, 13, 24, 85, 89-90, 91, 101-2, 109, 112, 178
cisterns, 8, 11, 20, 22, 28-9, 32, 37, 44, 49, 61-4, 75, 78, 83, 85-7, 107, 121, 123-4, 132, 171-2, 175, 180-1, 194, 197, 210, 217-8, 231-2, 235, 251, 255, 261
cisterns, See also *castellum aquae*
Civic councils, See municipal administration
Codex Iustinianus, 30, 46, 100, 257
Codex Theodosianus, 25, 30, 46, 100
Complutum, 59
Conimbriga, 41, 50, 156-60, **158, 159, 160**
Conimbriga - aqueduct, 39, 46, 159-60
Conimbriga - baths, 159-60
Conimbriga - *domus* of Cantaber, 73, 157, **161**
Constantinople, 12, 28, 32-3, 74, 131
Consuegra, 38
Córdoba, 41-2, 50, 75-6, 79, 80, **81**, 91, 101, 106, 118, 161-76
Córdoba - alcázar and *palatium*, 70, 80-1, 115, 117, 122, 162, 173-4, 175
Córdoba - aqueducts, 39, 41-2, 46, 76, 129, 163-76
Córdoba - aqueducts - Alcázar/*Qanat Amir*, 39, 55, 70, 115, 117, 173-5, **174**, 212
Córdoba - aqueducts - *Aqua Augusta*/Valedepuentes, 38, 48, 49, 88, 115, 163-7, **165**

Córdoba - aqueducts - Arroyo Pedroche/*Aqua Nova Domitiana*, 167, 190
Córdoba - aqueducts - Cercadilla, 114, 171-3, **172**
Córdoba - aqueducts - Western, 70, 76, 87, **114**, 114, 168-71, **169**
Córdoba - episcopal complex, 80-1, 90, 162
Córdoba - fountains, 70-1, 80, 114, 117, 164, 167-8, 170, 173-4, 175, 176
Córdoba - Great mosque, 70-1, 76, 117, **119**, 119-20, 163, 170, 175
Córdoba - sewers, 162
Córdoba - suburbs, 72, 75-6, 79, 80, 90, 163, 168-70, 175
Crete, 75
curiales, See municipal administration

dams, 81, 164, 178-9, 188, 191, 202, 211, 240, 241, 251
dams - dating controversy, 22, 194-7
drinking water, 7, 11, 56, 71, 74, 82, 148, 149, 194-5, 217

earthquakes, 26, 38, 48-9, 165, 171, 175
Ebusus, 38, 87
Egabrum, 37, 38
Egara, 90
El Tolmo de Minateda, 20, 62-3, **63-4**, 89, 102, 106
engineers, 46-7, 50-5, 101, 106, 221
episcopal complexes, 41, 46, 61-2, 64, 75, 77, 79, 80-1, 89, 110-1, 131, 142, 149, 150, 162, 172, 187, 200, 219, 223, 229, 235, 248

Feijoo Martínez, Santiago, 22, 194-5
Fernández Casado, Carlos, 17-8
Fez, 264
Fossa Carolina, 55
fountains, 11, 24-5, 44, 57, 59, 67-72, 78, 80, 92, 98, 117, 131, 155, 164, 167-8, 169, 170, 171, 173-4, 175, 176, 180-1, 199-200, 228, 230-1, 235, 245-6, 249, 255, 260
Frontinus, 1, 8, 15, 16, 30, 69, 194

garum and *cetariae*, 25-6, 40, 42, 137, 141-2, 150, 177, 180, 246, 255, 257-9, 260, 261
Gaul and Francia, 2, 7, 19, 25, 30, 32, 45-6, 51, 53, 90, 98, 106, 112, 130, 133, 141
Germanic peoples, 6, 13, 49
gradients and slopes, 10, 23, 32, 51, 159, 165, 208
Granada, 126
Guarrazar, 109

Huelva, 128
Hydatius, 15, 157, 160, 185, 186

Iberian Peninsula, 3-6, **4**
Ilipla, 38
inscriptions, 16, 25, 32, 33, 37, 68-9, 76, 83, 104, 105, 162, 164, 166, 167, 168, 176, 177, 187, 188, 191, 198, 212, 216, 228, 231, 233, 237, 242, 250, 251, 262
inverted siphon, 10, 37, 81, 139-40, 155, 167, 171, 227, 251
irrigation, 7, 11, 16, 29, 62, 87, 115, 128, 208, 210, 240, 241, 250
Isidore of Seville, 16, 73, 106
Islamic conquest, See Umayyad conquest
Islamic repairs, 22, 221, 234-5, 243-5, 248
Islamisation, 13-4
Italica, 37-8, 43, 84, 105
Italy, 2, 31, 45, 51, 72, 90, 104, 106, 110, 112, 130, 133

Jaén, 38, 126

John of Biclar, 15, 203
Justiniana Prima, 107

Lacipo, 38
Larache, 254
latrines, 27, 151
laws and legislation on water, 16, 29-30, 45-6, 48, 88, 100, 115-6, 118, 125, 240, 250, 251
León, 38, 149
Leptiminus, 28
Lex Visigothorum, 16, 241
Lisbon , 42, 50, 176-81, **177**, 220
Lisbon - aqueduct, 39, 46, **178, 179**
Lisbon - fountain, 68, **69**, 98, 180-1
Lisbon - Thermae Cassiorum, 177, 181
Liuvigild, King of the Visigoths, 38, 53-4, 104, 108, 203-4, 239
Lixus, 254, 255, 260, 262, 264, 265
Los Bañales, 38
Lucentum, 8, **9**
Lugo , 42, 50, 58, 75, 181-6, **181**
Lugo - aqueduct, 39, 46, 182-6, **182, 184**
Lugo - baths, 185

Madinat al-Zahra, 88, 115, 126, 164
Madrid, 127
Málaga, 56, 57, 60, 126
Marroquíes bajos, 7
Martos, 7
Masuna of Altava, 262
Medina Sidonia, 8
Mérida, 11, 42, 50, 56, 85, 90, 98, 105, 110, 118, 186-203, **186**
Mérida - alcazaba, **122**, 123, **124-5**, 188
Mérida - aqueduct repairs, 20-1, 43, 51-2, 104, 121, 198, 203
Mérida - aqueducts, 39, 42, 46, 129, 188-203, **189**
Mérida - aqueducts - Cornalvo, 88, 191-3, **192**, 203
Mérida - aqueducts - Las Abadías, 193-4, **193**
Mérida - aqueducts - Los Milagros/Proserpina, **21**, 51, **52**, 188-90, **190**, 201, 203
Mérida - aqueducts - San Lázaro, 190-1, **191**, 198, 202, 302
Mérida - baths, 59, 97, **100**, 200-1
Mérida - bridge, 46, 105
Mérida - fountains, 67, 195, 199-200
Mérida - Morería, 60-1, **61**, 73, 98, 122, 200
Mérida - walls, 46
Metz, 45, 90, 195
municipal administration, 12, 30-1, 45, 64-5, 72, 89, 93, 96-102, 110, 127

Narbonne, 32, 45, 51
Nîmes, 20, 32, 48, 72, 195
nymphaea, see fountains

Ocuri, 38
Ologicus, 89
opus signinum, 10, 22, 48, 53, 61, 63, 68, 82, 140, 145, 147, 155, 159, 164, 167, 171, 189, 194, 204, 208, 212, 215, 219, 227, 232, 245
Orense, 58-9
Ostrogoths, 103-4, 107
Oviedo, 55, 57, **58**, 109

Palma de Mallorca, 126, 127
Pamplona, 59, **60**, 87, 149
peer-polity interaction, 12
pipes, see water pipes
Pont du Gard, 10

Ravenna, 31, 45, 51, 54, 103, 131, 205
Reccared, king of the Visigoths, 62, 104

Recceswinth, king of the Visigoths, 57
Reccopolis, 19, 22, 27, 43, **72**, 89, 90, 101, 102, 106, 108, 203-10, **206**
Reccopolis - aqueduct, 39, 47, 53, 54-5, 65, 88, 104, 107-8, 114, 129, 206-10, **207, 209**
Reccopolis - fountain, 69, **71**, 205, 210
river water, 11
Romanitas, 11-2, 130-1
Rome (city), 2, 8, 12, 20, 24, 28, 31, 45, 49-50, 51, 54, 74, 87, 103, 131, 205
Ronda, 38, 126

Sagunto, 7, 88, 241, 250
Sala, 254, 255, 261, 265
Segovia , 43, 79, 109, 210-5
Segovia - aqueduct, 16-7, 20, 39, 46, 114, 208, 211-5, **211, 213, 214**
Septimania, See Gaul, Nîmes, and Narbonne
Seville, 43, 50, 75, 85, **86**, 105, 111, 118, 215-22, **218, 219**
Seville - aqueduct, 39, 115, 121, 129, 216-22, **217**
Seville - baths, 97, 216, 220-1
Seville - La Encarnación, 73, 218, **220**, 221
Seville - sewers, 28
sewers and drains, 11, 27-8, 44, 48, 62, 73-4, 88, 126, 149, 150, 162, 167, 175, 181, 185, 201, 205, 221, 233, 236, 237, 245, 246, 255
Sharia, See laws and legislation
Sidonius Apollinaris, 98
sieges and war, 49-50, 76, 153, 156, 160, 172-3, 178, 187, 197
sinter, 22, 23, 28, **47**, 47-8, 61, 134, 147, 151, 165-6, 215
siphon, See inverted siphon
specus, 10, 20, 47, 48, 53, 59, 79, 87, 145, 147, 151, 155, 159, 164, 166, 167, 171, 174, 179, 182, 183, 189, 197, 198, 202, 208, 212, 215, 232, 243, 258
spolia and recycled construction material, 22, 68, 125, 150, 168, 180, 200, 212, 246, 248
springs, 8, 11, 29, 30, 49, 56-9, 74, 80, 109, 112, 115, 126, 154, 159, 163, 165, 167, 169, 171, 183, 190, 191, 197, 207, 215, 230, 261, 265
springs - thermal waters, 58, 185
State formation, 6, 13
State formation - Idrisid, 262-4
State formation - Umayyad, 6, 13, 83, 89, 91, 113-9, 121, 131-2, 163, 263-4
State formation - Visigothic, 6, 12, 89, 98, 101-2, 104-8, 118, 204
substructio, 10, 49, 155
suburbs, 26, 46, 60, 61, 72, 75-83, 90, 97, 106, 122, 123, 144, 153, 153, 155, 163, 168-71, 175, 178, 187, 190, 200, 203, 219, 222-3, 229, 233, 237, 251, 252, 261, 262
Sueves, 6, 41, 42, 89, 153, 156, 158, 160, 176, 178, 182, 185, 187, 197, 202, 217, 222

Tamuda, 254, 260
Tangiers, 254, 255, 256, 261, 264, 265
Tarragona, 11, 43-4, 60, 76-7, **77**, 79, 80-1, 91, 101, 141, 222-36
Tarragona - aqueducts, 39, 46, 224-33, **227**, 235-6
Tarragona - aqueducts - Francolí, 208, 224-6, **225, 226**, 231, 235
Tarragona - aqueducts - Gayá, 80, 224, 224, 227-8, **228**, 235
Tarragona - aqueducts - Suburb, 229, 235
Tarragona - baths, 25, 97, 233, 235-6
Tarragona - cisterns, 28, 231-2, 235
Tarragona - *cuniculus*, 56, 80, 87, 229-30, **230**, 231

Tarragona - episcopal complex, 77, 90, 110, 111, 131, 219, 223, **231**, 231-2
Tarragona - fountains, 25, 56, 68, **70**, 98, 213, 228, 230, 231, 235
Tarragona - Francolí complex, 61, 75, 77, 80, 229
Tarragona - suburbs, 60, 76-7, 80, 90, 97, 223, 229-30, 233
taxes and taxation, 13, 30, 45, 99-101, 104, 117, 142, 204, 222
terminal cistern, See *castellum aquae*
Thamusida, 254, 255, 261, 265
Theoderic, king of the Ostrogoths, 45, 54, 103
Tiermes, 38, 87
Tingitania (Mauretania Tingitana), 7, 32, 39, 253-65
Tocolosida, 254
Toledo, 39, 81-3, **84**, 87, 90, 102, 105, 106, 115, 118, 126
Toledo - alcázar, 83, 115, 123
Toledo - suburb/Vega Baja, 46, 82, 106, 108, 123, 153
Tours, 90

Ucubi, 38
Umayyad conquest, 6, 13, 50, 55, 83, 117, 130, 154, 173, 176, 178, 186, 254, 261-2, 265
Umayyads in Syria and the Levant, 22, 33, 118-20
Umma, 113, 117
Urbanism - Islamic, 64, 70, 72, 76, 87, 121-7, 175, 239, 261
Urbanism - Late Antique, 89-93, 204-5, 223, 261
Urbanism - Roman, 2, 6-8, 11-2, 24, 36, 38, 74-84, 93
Uxama, 38

Valencia, 44, 59, 77-8, 79, 101, 236-49, **242**
Valencia - Almoina/Islamic conduit, 22, 115, **116**, 121, 248-9
Valencia - aqueduct, 39, 46, 88, 129, 149, 208, 239-49, **244**
Valencia - baths, 11
Valencia - episcopal complex/La Almoina, 46, 90, 110, 111, 131, 219, 235, 237-8, **238**, 245, 248
Valencia - Great well, 246-8
Valencia - *hammam*, 126, 248
Valencia - *nymphaeum*, 25, 67-8, 78, 131, 236, 237, 242, 245-6, **247**
Vandals, 106-7, 254, 257, 261, 262
venter, See inverted siphon
Victoriacum, 89, 106
Vienne, 30, 46
Visigoths, 6, 59, 76-7, 81, 82, 89-91, 98, 102, 103-9, 117, 130, 153, 156, 158, 162, 176, 182, 187, 197, 204, 239, 257
Visigoths, See also State formation - Visigothic
Vitruvius, 1, 8, 16, 96, 195
Volubilis, 255 262-5, **263**, 265

wafk, 113, 117
water mines, 41, 44, 80, 147-8, 155, 190, 217, 229
water pipes, 10, 15, 16, 22, 26, 27, 28, 37, 41, 48, 57, 61, 67, 68, 69, 73, 88, 139, 149, 150, 155, 159, 168, 171, 172, 178, 183, 195, 200, 202, 215, 216, 221, 236, 246, 251
waterproof stucco, 22, 166-7
wells, 8, 59-61, 73, 78, 80, 99, 124, 237, 246-8, 261-2
Wickham, Chris, 31

Zaragoza, 11, 44, 50, 99, 115, 118, 222, 249-51, **250**
Zaragoza - aqueduct, 39, 46, 250-1
Zaragoza - baths, 252
Zilil, 254, 256, 260

INDEX OF INSCRIPTIONS

CIL

II 175	176	
II 191	177	
II 2739	212	
II 2746	212	
II 2751	212	
II 4109	68, 231	
II 4509	145	
II² 14.1004	233	
II² 14.90	237	= ICERV 365
II² 7.217-9	69, 164, 167, 168	
II² 7.24	76	
II² 7.644	162	
II² 7.693	162	
II² 992	251	
VI 15258	92	
VIII 9835	262	
XII 4355	32	

ICERV

279	237	
303	99	
361	83	
363	187	
365	237	= CIL II² 14.90
366	104, 188	

www.ingramcontent.com/pod-product-compliance
Lightning Source LLC
LaVergne TN
LVHW011527200926
848484LV00001B/179

* 9 7 8 1 4 6 3 2 3 9 1 5 2 *